Under the general editorship of Harold D. Lasswell, Daniel Lerner, and Ithiel de Sola Pool.

The Emerging Elite: A Study of Political Leadership in Ceylon, Marshall R. Singer, 1964.

The Turkish Political Elite, Frederick W. Frey, 1965.

World Revolutionary Elites: Studies in Coercive Ideological Movements, Harold D. Lasswell and Daniel Lerner, editors, 1965.

Language of Politics: Studies in Quantitative Semantics, Harold D. Lasswell, Nathan Leites, and Associates, 1965 (reissue).

The General Inquirer: A Computer Approach to Content Analysis, Philip J. Stone, Dexter C. Dunphy, Marshall S. Smith, and Daniel M. Ogilvie, 1966.

Political Elites: A Select Computerized Bibliography, Carl Beck and J. Thomas McKechnie, 1968.

Force and Folly: Essays on Foreign Affairs and the History of Ideas, Hans Speier, 1969.

Quantitative Ecological Analysis in the Social Sciences, Mattei Dogan and Stein Rokkan, editors, 1969.

Euratlantica: Changing Perspectives of the European Elites, Daniel Lerner and Morton Gorden, 1969.

Revolution and Political Leadership: Algeria, 1954-1968, William B. Quandt, 1969.

The Prestige Press: A Comparative Study of Political Symbols, Ithiel de Sola Pool, 1970.

The Vanishing Peasant: Innovation and Change in French Agriculture, Henri Mendras, 1971.

Psychological Warfare against Nazi Germany: The Sykewar Campaign, D-Day to VE-Day, Daniel Lerner, 1971 (reissue).

Propaganda Technique in World War I, Harold D. Lasswell, 1971 (reissue).

Leader and Vanguard in Mass Society: A Study of Peronist Argentina, Jeane Kirkpatrick, 1971.

The Changing Party Elite in East Germany, Peter C. Ludz, 1972.

Technology and Civic Life: Making and Implementing Development Decisions, John D. Montgomery, 1974.

Scientists and Public Affairs, Albert H. Teich, editor, 1974.

SCIENTISTS AND PUBLIC AFFAIRS

Edited by Albert H. Teich.

The MIT Press Cambridge, Massachusetts, and London, England

Copyright © 1974 by
The Massachusetts Institute of Technology

This book was printed on Finch Textbook Offset
by Halliday Lithograph Corp.,
and bound in G.S.B. 5/535/34 "Deep Navy"
by The Colonial Press Inc.
in the United States of America.

Library of Congress Cataloging in Publication Data

Teich, Albert H.   comp.
  Scientists and public affairs.

  (M.I.T. studies in comparative politics)
  CONTENTS: Rich, D. Private government and professional science. —
Cahn, A. H. American scientists and the ABM: a case study in controversy. —
Nichols, D. The associational interest groups of American science. [etc.]
    1. Science and state—United States. 2. Scientists—United States. I. Title.
II. Series: M.I.T. studies in comparative politics series.
Q127.U6T38   353.008'55   73-22250
ISBN 0-262-20026-0

To the Memory of R. David Gillespie 1938-1972

CONTENTS

Harold D. Lasswell and Daniel Lerner

Not many responsible people deny that the makers and executors of public policy ought to utilize the available stock of pertinent knowledge and to anticipate the emerging needs of the body politic by arranging to obtain relevant knowledge well in advance of crises. Laymen and experts agree that these seemingly obvious criteria are frequently, if not typically, honored in the breach. The question that concerns us is whether we can explain the discrepancy between articulated norms and unsatisfactory levels of performance. The rich case material in the present volume can be fruitfully approached if certain general considerations are kept in mind.

Suppose we explore the implications of a hypothesis that refers to the elites of government, law, and politics in any body politic, whether the public order is characterized as "liberal," "socialist," "communist," "democratic," "authoritarian," "despotic," or any variation of these key political symbols. The hypothesis is that to the degree that political power is sought and is perceived to depend on the use of knowledge, knowledge will be sought and applied.

One is never entirely in error if he assumes that political power is at least among the preferred outcomes (the values) sought by those who engage in "the great game of politics." However, it is an empirical question in what measure power is pursued as an end in itself (a scope value), or is sought and used as a means (a base value) to other values, among which we identify all outcomes involving deferences and welfare.

Available knowledge of contemporary and earlier history points to at least one relevant finding: knowledge (that is to say, enlightenment) is rarely, if ever, a high-ranking scope value among decision elites. Of course, we often find that science and scholarship (or their equivalents in different cultures) are encouraged. They are supported as a base for further power, or for wealth and "glory" (respect), or for ethico-religious and other purposes.

Our hypothesis specifies that in the degree to which knowledge is perceived as facilitative of power, it will be encouraged and employed by the wielders of power. Who can doubt the overwhelming role of such perceptions in the backing given to nuclear science and technology, in the cultivation of astronomy by the imperial courts of Chaldaea-Assyria and ancient China, or in more recent centuries by the British Navy, or in our own time?

These are famous instances of the affirmative encouragement and use of knowledge. They can be offset, if not overshadowed, by the notorious failure of decision processing at every level (national, subnational, or transnational) either to mobilize pertinent information from the stock of current knowledge or to provide the assets required to foster the growth and dissemination of the cultural, biological, and physical arts and sciences. Recall, in this connection, the belated attack on the problems currently subsumed under the label of "environment."

It will be noted that many of the problems relating to the environment are not narrowly focused on the exercise of coercive authority. That is to say, these problems do not "directly" involve political power. They do, however, interest power specialists to the degree that probable success or failure in the arenas of external or internal politics is perceived as depending on environmental action.

A further point is that participants in the decision process are never exclusively concerned with the pursuit of power for themselves as independent persons or for the groups—national, religious, scientific, or what not—with which they identify the "self." To the extent that pollution is perceived as a threat to health, for instance, knowledge of the factors that affect pollution is sought in the degree to which well-being is a priority value in the power system that allocates all values.

The original hypothesis needs to be interpreted as involving more than the value priorities of effective decision elites. Included are commitments to patterns of value distribution as inclusive or exclusive of the members of a body politic. Where popular government prevails the commitments are to broad rather than narrow sharing of power, well-being, wealth, and all other values.

An implication of the analysis is that, in any society, the use of knowledge is heavily conditioned by the socialization of all effective participants in the decision process. Where values are plural, and knowledge has a high priority, the chances are that the formal and informal institutions of decision foster the institutions of knowledge.

A further implication is that, in a pluralized society, those who follow science and learning as a career will themselves be sufficiently pluralistic in value orientation to integrate demands on behalf of "science for science" with "science for human fulfilment." To what extent can the actual state of affairs be discovered? Currently available are many instruments for appraising the degree to which the "knowledge community" in the United States, for example, conforms to the model appropriate to a pluralized society. Although proper procedures are at hand, it is not yet customary for professional societies, universities, or other knowledge institutions to conduct regular self-appraisals, or to encourage systematic appraisal by others of the use of knowledge in the decision process, including knowledge of the process itself. Perhaps the materials in this volume will hasten the evolution of a network of structures that are presently in an underdeveloped state.

It is appropriate that these materials should be drawn from Ph.D. theses in the "Science and Policy" program at M.I.T., which has taken vigorous leadership in translating the ancient contest between Knowledge and Power into terms appropriate for our contemporary world situation. We welcome this book into our MIT Press Series with confidence that it has much to teach those who would learn.

Harold D. Lasswell
Yale School of Law, Yale University

Daniel Lerner
Massachusetts Institute of Technology

Harvey M. Sapolsky

One popular image of the scientist is that of a lone researcher mixing compounds through the hours of the night in the cluttered basement laboratory of his home. The solitude of creative work notwithstanding, the scientist is in actuality an organizational man. The support for his research more frequently than not comes from an agency of the government or a large corporation. He receives his training within the facilities of the nation's large universities. And his work goes on in laboratories tended by teams of assistants and technicians. No one stands alone in science today.

An integral part of the organizational world of a scientist is supplied by the scientific societies. These organizations, which link scientists of similar inter-ests, provide opportunities for professional recognition and intellectual discourse within science. Though voluntary associations, the scientific societies offer services so vital that no scientist can avoid membership in at least one of them. Their dues are in this sense as compulsory as the taxes of public government and are recognized as such by the public tax collectors.

In the following essay Daniel Rich examines the functions and governance of the scientific society. His perspective is that of a political scientist, for he sees the scientific society as an organization that attempts to allocate values authoritatively within a community. Delineated in the essay are issues such as the establishment of professional standards, the promotion of the economic welfare of the members, and the legitimization of managerial authority that have come to dominate the organ-izational agenda of scientific societies.

Rich's analysis of society governance is particularly penetrating. Power within the societies, he points out, naturally gravitates to the executive directors because they devote full time to societal affairs and control the internal communications of the societies. The executive directors' exercise of power, however, must always be cautious since they are the nominal subordinates of elected officials and since they are vulnerable to the countercurrents of meritocracy and democracy that comprise the political ideologies of the scientific societies. The precariousness of their position within the societies leads the executive directors to be conscious of their management skills and to form professional societies of their own. Ironically, as Rich notes, those that manage scientific societies have come to identify with a profession different from those they serve.

Rich also considers the opportunities for cooperation, and especially for consolidation, among scientific societies. Although he describes successful efforts
to merge and to federate groups within several fields, Rich sees the government
of science remaining pluralistic. In his view, progress in science spawns new
societies by encouraging the fragmentation of existing fields of study and the establishment of new ones. Scientific societies thus created are necessarily rivals
in a continuing struggle for the affiliations of scientists.

From a science and public policy perspective, however, the essay's most interesting section is that dealing with the societies' relations with the federal government. Despite their potential for aggregating and articulating the interests of
scientists, the societies have engaged in relatively few attempts to influence public
policy. Historically, this self-restraint can be explained in terms of both the
federal government's obvious indifference toward science and the scientists' own
fears of the consequences for science of a direct involvement in "politics." More
recently, as the federal government came to promote research activities, there
was apparently little need to protect the interests of science since they were well
served by the growth of science-oriented agencies within the government.

Self-restraint need not be permanent. The current decline in the nation's willingness to promote science could alter both the governance of science and the relations
between scientific societies and the federal government. As Rich notes, a debate
is developing within the societies over what should be their role in aiding the economic welfare of members and in guiding the science policies of the federal government, two highly related issues. New organizations are forming to consider these
problems, and the leadership of the established societies is seeking ways to demonstrate its concern. The fear of loss, it seems, is a greater stimulant for
political activity than is the anticipation of gain.

How far the societies actually move in the direction of interest aggregation and
articulation could well depend as much upon the judgment of federal officials
sympathetic to their cause as upon pressures from their members. In other policy
areas it has often been the committed legislator or the professional within the bureaucracy who has held the initiative in coordinating disparate elements into a cohesive
source of lobbying support. Their assessments of the political opportunities in a
hostile environment are difficult to ignore. If the friends of science in public government favor a realignment in the internal government of science to meet the needs of
politics, the obstacles to cooperative action among the scientific societies may begin
to fade.

Daniel Rich

INTRODUCTION

In universities, corporations, churches, and a host of other private associations, the legitimacy of present structures of decision making and of the decisions made is being questioned by some of those individuals affected. In each case there are efforts to reevaluate the responsibilities of the institution and the structures deemed appropriate to fulfill such responsibilities; in each case there are issues and conflicts not generally associated with private institutions and thought to be uniquely characteristic of public government. This turbulence has served to make manifest political aspects of institutional life which formerly were submerged; it has focused attention on the political nature of institutions which were believed to be nonpolitical.

Nearly thirty years ago Charles Merriam articulated the need for political scientists to overcome the conceptual bias of their discipline; to consider political processes in a wider range of settings. Nevertheless, research on private government, on politics and governance in private institutions, has been sporadic and noncumulative. Merriam observed that while political scientists in private spoke of "university politics" and "office politics," in their professional lives they neglected private government. [1] It appears as if their experience led them to recognize political processes wherever they took place while their professional roles rendered much of that insight either illegitimate or inconsequential.

This analysis presents a profile of contemporary scientific societies as private governments. [2] Based upon case studies constructed from interviews, document analysis, and personal observation, it seeks to clarify [3] (1) the nature and implications of their functions, (2) the predominant characteristics of their internal government, (3) the nature of relationships among these institutions, and (4) the relationship of these private governments to public government.

Contemporary scientific and technical societies have come to resemble professional associations in such fields as medicine and law. Like such associations they operate as private governments; they conduct programs and offer services "that might otherwise have been done through more private enterprise or not done at all..." and they "regulate the behavior of their members—their citizens." [4] Generally they "represent the interests of their constituency in relationships with other groups or governments." [5] They are instruments of social control and vehicles for the promotion and protection of professional autonomy; they authoritatively allocate values within the fields they serve.

At the same time, however, in comparison with professional associations in medicine and law, scientific societies are weak private governments. None has achieved the nearly monopolistic control of the AMA for example. In some fields multiple societies competing for preeminence have perpetuated a fragmentation of authority. Moreover, as systems of professional control, the scientific societies have shared authority with other scientific institutions, particularly the universities.

The operation of scientific societies as private governments has posed serious questions of legitimacy. Contemporary societies have evoked a complex array of functions and clientele. As a result there appears much confusion and uncertainty concerning what these institutions should do and whom they should serve.

PRIVATE GOVERNMENT AND PROFESSIONAL AUTONOMY

Diversity and complexity are the most outstanding features of contemporary scientific societies as private governments: they aim to serve a variety of purposes by performing a multiplicity of functions, some of which are seemingly inconsistent; they exhibit a fragmented structure of government with authority distributed among both formal and informal bodies, they manifiest an inconsistency in their relationships with other private governments, at one time cooperative and another time competitive; and they relate to public government in some ways intended to enhance their autonomy and in other ways intended to more deeply immerse them in society.

In large measure the functions of contemporary scientific societies have developed as adaptations to the professionalization of science. Professionalization began in the nineteenth century with the formalization of scientific careers and the specialization of scientific fields. It has been a gradual and halting process, developing at different rates in different fields. It has often met with resistance within the scientific community; nevertheless, professionalization continues to be a dominant influence on scientific institutions.

Scientific fields have become professionalized as they have acquired status as occupations requiring specialized knowledge and characterized by a service or collective orientation, that is, an awareness of public image and public responsibility and a determination to be respectable and respected.[6] Manifestly and latently, the scientific societies have contributed to the professionalization of their fields.

Initially, they promoted the preemption of scientific skills;[7] the process by which "a task, which has customarily been performed by one group or by everybody in general, comes into exclusive possession of another particular group."[8] Through the preemption of scientific skills, the privilege of practice becomes restricted to those with specialized training.

By providing for scientific interaction and discourse, the scientific societies
have also contributed to the development of a collective consciousness of status
such that individuals carrying on work in scientific fields may develop a sense of
common identity. As a result, individuals come to consider others performing
similar tasks as a reference group, believing that they share a common destiny
with other practitioners.

Professional status requires public affirmation. William Goode has suggested
that the "advantages enjoyed by professionals... rest on evaluations made by the
larger society, for the professional community could not grant these advantages to
itself. "[9] By encouraging the diffusion of knowledge and by promoting public acceptance of special status, the scientific societies have fostered both scientific and
public legitimacy.

By contributing to the preemption of scientific skills, the collective consciousness of status, and the achievement of public and scientific legitimacy, the scientific societies have helped to ensure the professional status of their fields. As professional status has been achieved, the societies have reduced their efforts. Attention has been directed at the attainment of professional autonomy; concern is focused
on governing the institutionalized profession. The autonomous profession exercises
exclusive and decisive control over the professional lives of its members. The
scientific societies have become instruments of professional autonomy. They help
to control professional socialization, to enforce norms of practice, and to regulate
relations between the profession and society.

As they have contributed to professionalization and particularly as they have served as instruments of professional autonomy, the scientific societies have become
more influential private governments.

Through publication of journals and sponsorship of meetings and awards, scientific societies continue to maintain and reinforce preemption, consciousness of status,
and legitimacy. Such activities also provide a degree of control over the distribution
of status and rewards within the profession. Through a variety of programs ranging
from group insurance to collective bargaining, scientific societies influence the
economic welfare of their members and consequently intensify the attachment of
their members to the society. By means of standards of admission, codes of ethics,
licensing, and so on, they exercise control over professional socialization and practice. Acting as spokesmen, they influence the relationship between the professional
and society, particularly public government.

Publications, Meetings, and Awards

Contemporary scientific societies, like their earliest counterparts, foster scientific communication through the publication of journals and the sponsorship of meetings. Publications and meetings reinforce professional status. As areas for the diffusion of specialized knowledge and the recognition of expertise, they underscore the preemption of scientific skills. In addition, as publications and meetings provide for interaction and common communication they promote collective consciousness of status. Moreover, through specialized publications, such as news magazines or through sections of more technical journals, news and information about the profession are transmitted which reinforce mutual identification and diffuse common professional values. Publications and meetings also contribute to the public legitimacy of the profession and often gain publicity for it.

While meetings and publications contribute to preemption, status, and legitimacy, they also provide scientific societies with opportunities to exercise control over the distribution of rewards in scientific fields. Publication as an indicator of research productivity remains an important criterion for achieving status in the scientific profession.[10] Acting through editors or editorial boards, scientific societies influence the distribution of status and the tangible economic rewards accruing to status. In deciding which research will be published and which papers will be presented at meetings, the societies operate as vehicles for professional recognition and provide criteria for the evaluation of professional performance. "The scientific community's major means of control over its members' standards of work resides in its capacity to grant or withhold recognition of their work."[11]

Most societies annually make awards for outstanding accomplishment in their fields. Such awards may buttress group identity and establish "folk heroes" for the profession.[12] They also foster public recognition of professional competence and contribute to the "selling" of the profession to the public. Symbolically, awards take on importance as they are indicators of the society's capacity for control over individual recognition and professional achievement. In this sense awards are not only important for the individual scientist but "they also have a vital function for science as a whole."[13]

Welfare Functions

Particularly since World War II, scientific societies have been increasingly involved in the economic lives of their members. Beyond the indirect influences exercised through publications, meetings, and awards, societies offer services ranging from group insurance and pension plans to occasional involvement in collective bargaining.

In the early stages of development some societies occasionally engaged in lobbying activities on behalf of their members. For the most part, however, contemporary societies have avoided economic lobbying and resisted the development of an image as a trade association or union.

Placement services, generally are regarded as of secondary importance, have occasionally developed into an important activity when some perceived economic or employment crisis generates demand for economic services. For example, the American Meteorological Society became actively engaged in employment services immediately after World War II, when it appeared that because of reductions in military needs there would be insufficient job opportunities for meteorologists trained during the war. The society then launched a program of reorganization and professionalization designed specifically to meet the needs of demobilization and to establish a footing for meteorological services in the private sector of the economy. Particular attention was paid to employment services such as counseling and job placement.

More recently, leaders of various societies have reported increasing pressure from some of their members to expand their economic services. This pressure has resulted largely from currently rising unemployment rates in scientific and technical fields. For example, leaders of the American Institute of Aeronautics and Astronautics report that members are looking to their society to provide help in coping with unemployment resulting from cutbacks in the aerospace industry. On the West Coast, in particular, efforts have been made to improve counseling and placement activities. In addition, members have looked to the institute to protect their employment status through other activities such as sponsorship of portable pension plans.[14]

A similar pattern of response to perceived economic crisis is evident in the field of physics. The American Institute of Physics has for many years offered members employment services, but these services have not been perceived as particularly important. Recent high rates of unemployment have led some members of the institute to demand that greater attention be given to their economic security. Indeed, dissident members established a separate society, the American Physicists Association (APA), arguing that the American Institute of Physics either lacks the desire or the resources to represent adequately the economic interests of physicists. The APA was created in the spring of 1969 by a small group of young physicists at the University of Maryland expressly for the purpose of translating "concern about the economic problems facing physicists into concrete, positive, and constructive action."[15] Its first goal was to develop a large membership and organizational base.

Its more substantive and long-range goals were (1) "to bring into balance the supply and demand of physicists," (2) "to make a concerted effort to present to Congress and to the general public the ... values of physics ...," (3) "... to employ profes - sional lobbyists to work in Washington, D.C. for the economic interests of physi- cists," and (4) "to help interested members find desirable positions...."[16] One of the effects of the APA's existence has been to put pressure on the AIP to expand its own economic services. Some leaders of the AIP indicate that the existence of a "competitor" may result in the institute becoming more active in behalf of the economic interests of its members. They are, however, reluctant to become in- volved as an economic lobby since this activity would threaten the tax-exempt status of the institute. Nevertheless, the existence of the APA has led the AIP to reevaluate and further emphasize its placement services.

In addition to employment services, some scientific societies offer members group insurance and medical plans as well as pension and retirement plans. More- over, "several societies are applying limited economic pressure on employers by publishing surveys of salaries of scientists and engineers; as a result, firms pay- ing substandard rates face the threat of losing professional employees and of find- ing it more difficult to recruit new ones."[17] Such efforts, of course, become less effective when there is a scarcity of jobs for practitioners.

Some societies also have attempted to influence the employment relationship. Thus, leaders of the American Meteorological Society occasionally have intervened on behalf of their members with regard to employment conditions in the Weather Bureau, but such intervention has never taken the form of overt lobbying. Generally, scientific societies have rejected collective bargaining. Surveys of membership attitudes indicate demand for greater attention to this and other economic activities. For example, the Strauss and Rainwater survey shows that most chemists feel they do not have sufficient status on the job and wish that their society were more active in their behalf.[18] "Yet, even though there is considerable demand for actions that closely resemble those traditionally undertaken by unions, all surveys of the mem- bers of ... societies report overwhelming sentiment against unions and collective bargaining."[19] Such efforts have generally been left to what are portrayed by Wil- liam Kornhauser as professional unions—groups such as the Engineers and Scien- tists of America or the American Physicists Association.[20]

Controlling Professional Socialization and Practice
By means of standards of admission, codes of ethics, and licensing, scientific soci- eties exercise control over professional socialization and practice. In this way, scientific societies have reinforced professional autonomy.

Most scientific societies are influential in professional education; they perform a wide range of services including student recruitment, counseling, guidance, establishment of curricula guides and plans, and sponsorship of student branches of local sections. For example, the AIP provides counseling services to high school students who are selecting colleges to attend, publishes and disseminates information about the profession, reviews and evaluates curricula in high schools and universities, sponsors a visiting scientist program, and acts as an "umbrella" under which the Society of Physics Students operates. Such activities contribute to the institute's control over admission into the profession.

Admission to most scientific societies has become more restrictive as these institutions have become more professionalized. In most cases societies distinguish among grades of membership based on professional accomplishment. The lower grades often continue to be open to virtually anyone demonstrating an interest in the field. Most frequently, however, only members at higher grades acquire full service and voting privileges as well as the right to hold office in the society. Thus, the American Meteorological Society, which was established without restrictions on admission, has become more selective in its standards. One of the society's officers has stated:

The AMS is becoming more and more professionalized. We are dropping the amateur status. There is more and more specialization of meteorological careers. The drift is in the direction of the AMS becoming more like the AMA or the ABA. It may be that in the future the society will drop all but the professionally trained members. The amateur would be taken care of through a magazine... but not membership. [21]

The grades of membership are intended to be a status hierarchy. Upward mobility from one grade to another is an indication of increased professional accomplishment and recognition, a sign of prestige. In most societies an upper grade of honorary member or fellow is reserved as an award for exceptional service to the profession.

As societies have attempted to restrict admission, they also have attempted to restrict competition within the profession. Competition is anathema to professionalism; it is counterproductive to efforts by the profession to "preserve a solid front vis-à-vis the public in general"[22] and to reinforce what Seymour Martin Lipset has identified as the "cult of Unity."[23] Unity is believed necessary to safeguard a public image of expertise and "unity requires that competition be curtailed."[24] In this context a number of societies have established (or have considered the establishment of) codes of ethics to restrict competition. The AMS, for example, has created a code and established a board of professional ethics to investigate alleged violations.

The code specifically directs attention to the development of a norm of noncompetition; it admonishes meteorologists to avoid competitive practice and demands that they not discredit other meteorologists. In addition, the code emphasizes the collective responsibilities of meteorologists with regard to clients and the general public; it demands that members of the profession discourage sensationalism, base practice on sound scientific principles, and direct activities away from practices generally recognized as being detrimental to or incompatible with the general public welfare. Implied is that conformance to the code will further justify the profession's claims to special status.

In practice, the AMS code rarely has been enforced. In those few cases when charges have been made, the actions of the board of review have been limited to warnings or admonitions.

Enforcement has been made even more difficult since some members, those working in the private sector, object to the existence of the code. The emphasis on noncompetition is viewed as unrealistic and counterproductive to the fortunes of private meteorology.[25] Such objections have served as an obstacle to establishment of codes of ethics in other fields. For example, the board of directors of the AIAA—many of whose members are prominent in the aerospace industry—appears reluctant to institute such a code. However, some AIAA leaders believe that the majority of the membership favors a code and that it will be established over the board's objections.[26]

In the past, most societies avoided licensing and certification of practitioners, since such activities were thought to be functions appropriate for engineering but not scientific societies. Nevertheless, the American Chemical Society supports two national certification programs for clinical chemists and has worked with the Council of State Governments on a model bill to regulate clinical laboratories and their personnel. Similarly, the American Meteorological Society conducts a certification program for consulting meteorologists and a program of evaluating and distributing seals of approval to radio and television meteorologists. Such programs may be intended to regulate the supply as well as the quality of meteorological work. Moreover, as E. C. Hughes points out, through licensing and certification "competence becomes an attribute of the profession as a whole rather than of individuals as such."[27]

Consequently, the public may be protected, but the profession is also protected by the fiction that all licensed professionals are competent and ethical until found otherwise by their peers."[28] The preemption of skills becomes complete.

Public Relations

Scientific societies exercise control over the relationship of their professions to
society at large. In part this control is apparent in the enforcement of codes of
ethics and in licensing procedures. It is also apparent insofar as publications and
meetings contribute to public and scientific legitimacy for the profession. In addition, societies perform activities that are explicitly intended to promote and protect the public status of the profession.

Leaders of various societies appear convinced that their institutions should act
as professional spokesmen. Some indicate that they hope that the public will associate the society with its field in a way similar to the conventional association of the
medical profession with the American Medical Association. Thus, the executive
director of the AIAA indicates that, while the constitution defines the institute's
role narrowly, the AIAA should "provide leadership for the aerospace profession..."
and that "in the future the AIAA will look more like the AMA in the sense that it
will provide a powerful voice for getting the profession heard."[29] Other leaders of
the institute articulate similar sentiments. One indicates that the institute "has the
job of making members believe they are one group with one interest."[30] Another
suggests that the AIAA should concern itself more aggressively with the profession's
welfare and become more active it its public affairs.[31] Toward this end the institute
has sponsored public forums, debates, and position papers on particular issues.
In addition, it supports a Washington staff to keep the institute informed of governmental policies affecting the profession. Other societies engage in similar activities.
There is, moreover, evidence that at least in some societies the memberships desire
that their institutions become more active spokesmen and promote the public status
of the profession. Thus, a study of the American Chemical Society indicates that
most chemists feel that the general public image of chemistry is low and that the
American Chemical Society should be more active in promoting a positive image
for the profession.[32]

To support their role as spokesmen and to facilitate their institutional representation of the profession, scientific societies such as the AIAA and the AIP employ
public relations personnel to provide information to the media and promote public
understanding and consciousness of the profession. The AIAA's former director of
public affairs provides the following commentary on the function of his department:

The AIAA is kind of the AMA of the aerospace profession... my job is to woo the
public... to gain public support for aerospace science and engineering. But selling
the AIAA is different from selling a product—we are too vulnerable to bad publicity.
If I tried the "hard sell" the institute would be in trouble.[33]

Public relations men act as buffers between the profession and the public, or more precisely between the profession and the media. They attempt to stimulate a positive public image for the profession by channeling favorable information to the media and usually, with less success, restricting criticism. The latter function generally is dependent upon personal relationships between the P.R. men and representatives of the media. The former function is more actively pursued. Thus the public relations director of the AIP proposes:

Our job is to provide better public understanding of physics and astronomy and of physicists and astronomers; to enhance the attractiveness of these fields; to underline the role of physics in strengthening the national economy; to demonstrate the contribution of physics to our stockpile of knowledge and to our culture. [34]

Public relations representatives do not seem optimistic about the effectiveness of their efforts to enhance the public status of their professions. Moreover, as one P.R. man points out, public relations does not really influence the public directly; "the most we can do is provide reliable information to the media."[35] The job of "wooing" the public is then really a job of "wooing" the nation's science writers and commentators. Thus, the AIP director of public relations suggests: "We look upon the nation's science writers as multipliers of physics information; they are our primary audience."[36] In this context, the sponsoring of science writer seminars and the conferring of awards for excellence in science writing are significant activities.

Nevertheless, the molding of a public image remains a difficult, if not impossible, task. As Corrine Gilb points out, a profession's public image "largely reflects the public's reaction to thousands of individual contacts; or is the combined product of many portraits in motion pictures, plays, novels, magazine stories and the newspapers."[37] Moreover, the public image of the profession, accurate or not, has the effect of a self-fulfilling prophecy. That is, the image conveyed to the public "undoubtedly influences certain personality types to enter the profession and may even affect professional men's behavior": the image "helps to create the fact."[38] Reflecting on the difficulty of his task, one P.R. director points out that ultimately the only way of really affecting the public image of his profession is to make every practitioner his own P.R. man.[39] Such attitudes reemphasize efforts at professional socialization and standardization of practice through codes of ethics and licensing.

Contemporary societies engage in activities that suggest a growing resemblance to modern professional associations. Like these associations they foster professional autonomy and serve as vehicles for the governance of the autonomous profession.

As a result, scientific societies exercise control over the professional lives of their members and over the relationship of the profession to society.

At the same time, however, in comparison with other professional associations, scientific societies remain weak private governments. They have become involved in the economic lives of their members, in the enforcement of professional standards, and in the control of the profession's relationship with society. But, to date, all of these involvements have been marginal and few have demonstrated effectiveness.

Moreover, while the operation of scientific societies as private governments has received support from some leaders and members, it has also encountered resistance. Some argue that the sole purpose of scientific societies ought to be the advancement of science through the performance of traditional functions. It is believed that professionalization represents an unjustified distortion of the underlying purpose of scientific institutions, that it represents a commitment to the career interests of members or to the association itself at the expense of the advancement and dissemination of scientific knowledge for its own sake. Such attitudes are a continuing source of resistance to the growing influence of scientific societies as private governments. Moreover, even those who advocate a more influential role for scientific societies often appear uncertain about the implications of such a role. Important questions seem unresolved. Should professionalization proceed as it has in fields like medicine and law or is there something unique about science as a profession? Given that scientific societies should no longer be singularly dedicated to the achievement of purely intellectual goals, then to what additional goals should they be committed?

THE POLITICS OF PRIVATE GOVERNMENT

As scientific societies have become more active vehicles of professionalization, the structure of government within these institutions has been modified. Influence in scientific societies has gravitated to a small set of paid employees headed by an appointed executive director. The influence of this group is predicated upon the possession of managerial skills as well as their proximity to the daily affairs of the institution. Thus, they constitute a skill elite that operates within a governmental structure best described as a limited managerial oligarchy in which managers share influence with elected officials and technical committees.

The appearance of oligarchy in scientific societies is not unexpected. Indeed if we are mindful of Michels' "iron law of oligarchy," then only the absence of oligarchy in an organization should be surprising. Moreover, as Philip Selznick points out, most voluntary organizations are skeletal, manned by a small core of individuals, with the membership performing a marginal role in decision making. [40]

To suggest that oligarchic tendencies are expected is not also to suggest that such tendencies are without significance. The appearance of oligarchy is especially significant if it occurs in institutions which claim to be nonoligarchic. Indeed, as Seymour Martin Lipset points out,

Oligarchy becomes a problem only in organizations which assume as part of their value system the absence of oligarchy...when one finds an organization ostensibly devoted to the extension of democracy which is nevertheless itself undemocratically governed, some explanation seems demanded. [41]

Generally scientific societies do not claim to be democratic institutions. Rather they tend to present themselves as representative meritocracies; institutions in which leaders are held accountable to constituent members and in which ascendency to positions of leadership is related to scientific achievement. Given such pretentions, the appearance of oligarchy in scientific societies requires some explanation.

Formal Patterns of Authority

Generally, formal authority to govern scientific societies is vested in an elected council and president. The constitutions of the majority of societies specify that the council is to operate as a legislative body and that the president is to serve as chief executive official. The council's responsibility is to make policy so as to exercise general charge of the affairs of the society. Thus, the constitution of the American Meteorological Society provides the following mandate to the council:

The Council is the principal governing body of the Society....It is responsible for ensuring that every reasonable action is taken to accomplish the purposes of the Society. [42]

In contrast, the president is to act as the "executive arm of the Council" to see that decisions are effectively implemented. [43]

Constitutionally, the form of government in scientific societies is like that common to other voluntary associations, particularly to other professional associations. Moreover, there is a striking resemblance between the governmental forms of professional associations and the structure of American public government. As Corinne Gilb points out:

...the professional associations have written constitutions, though only limited judicial review. They have representative government, universal suffrage for all members, secret ballot, parliamentary procedures.... Their legislatures, which are often larger than American public legislatures but meet for briefer durations, resemble public legislatures in some of their forms and procedures. There are similar problems in determining the basis and manner of apportionment and representation; many of the private legislatures have speakers and standing and interim committees like those of public legislatures; state delegates to the national private legislatures caucus as state delegations do in the federal House of Representatives; and "programs" come down from above to private legislatures, as it does from the executive branch to public legislatures. [44]

As in public government, the apparent separation of powers between legislative and executive bodies tends to break down in practice. In addition to their other duties, presidents of most societies serve as chairmen of executive committees. Such committees, generally composed of council members and various executive officials in addition to the president, operate as a surrogate for the legislative council. Like the boards of directors of large corporations, these executive committees may virtually co-opt the authority to run the institutions "because the assembly of members or their representatives meets so briefly and infrequently (at most twice a year) and because delegates to the assembly are farflung. "[45] They may perform executive and judicial functions in addition to their legislative operations. Moreover, their authority is often formally sanctioned under the society's constitution. This structural concentration of authority leads Gilb to conclude that "Montesquieu's doctrine of separation of powers which had so much influence on American public government has not significantly affected the government of American professional associations. "[46]

Generally, executive and legislative officials are elected by the membership. Lacking party systems, nominations are made by appointed committees frequently consisting of past presidents. [47] Despite universal suffrage, voting participation is low. [48] Like most voluntary associations, the membership plays only a peripheral role in governance. One AIAA official points out, "The members are largely apathetic; they are interested in their own careers and problems, not in the profession as a whole. "[49]

Although elected officials possess formal authority, they have only limited influence over the day-to-day affairs of their societies. Elected leaders have short tenure, and not infrequently an official is replaced before he becomes truly familiar with the society's activities. There are few material incentives for elected officials to devote a great deal of their time to their societies. Usually they receive no financial compensation and must pay their own expenses. Invariably they are employed in jobs demanding a sizable time commitment. Moreover, they are often geographically isolated from one another and from the society's headquarters. Councils meet

together as governing bodies only a few times a year; executive committees meet
more often but generally not more than a half dozen times a year. In addition,
presidents may be away from headquarters for long periods of time.

Rewards for holding office in scientific societies are largely honorary. Moreover,
elected officials gain whatever prestige an office carries regardless of the time
they devote to their positions. They are usually nominated because at some point
in their career they demonstrate significant scientific achievement, not because
they possess a capacity for governance or administration. In most cases they cannot
be reelected to office; thus, their past records are of little consequence to their
future status in the society. Indeed, members of these societies do not seem to
expect that officers will devote singular or even primary attention to the governance
of their institutions. One member suggests that "most presidents are content to be
passive and to carry out largely ceremonial functions." Another admits that "these
are busy men; we cannot expect them to ignore their own work to run the society."

Since opportunities and incentives to exercise influence are limited, there is a
tendency for elected officials to devote their attention to a few of the most salient
issues facing their institutions. According to one executive director, "most presidents when they come to office, pick out one or two matters of particular interest
and work principally on those." Similarly, an official of the AMS suggests that most
presidents "are happy to work on a few big issues and stay out of the business of
running the society." On particularly salient matters these officials may exercise
decisive influence but on most other matters they defer to others.

Managerial Oligarchy

In their early histories, when memberships were small and resources and programs
limited, influence in scientific societies tended to gravitate to a single individual or
small group. Not infrequently, the founder or founders retained significant control.
Charles Brooks served for over twenty years as secretary of the AMS. For most
of that period the headquarters were in his home and office, moving with him when
he changed research positions. Brooks edited the Bulletin, wrote its editorials, and
took charge of general administration. Current leaders identify him as the single
most influential individual in the early history of the AMS; indeed, he has achieved
the status of a folk hero within the profession.

Similar patterns are evident in the development of other societies. Karl Darrow
served as secretary of the American Physical Society for nearly thirty years, and
Lester Gardner and Jerome C. Hunsaker, founders of the IAS, remained influential

from the institute's inception through World War II. Little in the way of policy was enacted without consulting these men. In each case they remained influential by accepting the responsibility for administering the daily affairs of their societies.

As the fields they serve have professionalized, scientific societies have become complex organizations, with tens of thousands of members, budgets in the millions of dollars, and a variety of comprehensive programs. Complexity has necessitated the growth of bureaucracies to provide managerial expertise.[50] "Managing a scientific society is like managing a big business; a professional bureaucracy is necessary to keep it running smoothly."

Thus, the executive director has emerged as the most influential individual in scientific societies. He is the primary managerial expert and the head of a managerial staff which provides significant resources to reinforce his own expertise. Moreover, he tends to have long tenure in office which contributes to the further development of his expertise.[51]

He is the pivot wheel in the professional association and probably its most powerful individual, though nominally he takes his directions from other people. He is the expert among amateurs, for he usually holds his post a number of years, whereas association officers and board members come and go.[52]

In most organizations scientists are experts and managers are clients. In scientific societies this relationship is inverted. Managers gain power because they possess expertise; they serve as a skill elite for these private governments. "As problems become more technical and complex the amateur and transitory officers of associations more than ever must put their faith in staff men to map out programs and make the daily decisions."[53]

Formally, the executive director and his staff are empowered only with authority to implement policies of the council and president. In fact, as in the case of other bureaucracies, the authority to implement carries with it the authority to legislate. Elected officials may retain control over the broad outlines of policy but the managerial staff controls the execution of policy and the daily affairs of the institution. One executive director points out:

Anyone who carries out policy is also in a strong position to influence policy. Some of my colleagues view their positions more passively than I do. I see my job as being an activist, of promoting ways in which the society can and should change. But in any case you can't help influencing policy.

The position of the executive director, his day-to-day presence, renders him influential and provides continuity in the work of the society. One incumbent concludes that "if you want things done you need an executive director who is willing to accept and exercise influence."

Executive directors do not possess legitimacy to legislate policy. As a result their influence must be applied through covert channels. According to one executive director,

People expect us to stay in the background. I have more influence than most people think but to maintain my position I must exercise that influence carefully, behind the scenes, so as not to appear to run things.

Not infrequently, executive directors attempt to co-opt the legitimacy of those who possess formal authority. Co-optation is particularly evident in the relationships between executive directors and presidents. Presidents rely on executive directors to execute policy in their name, and executive directors depend on presidents to legitimize their own influence. Not surprisingly, executive directors identify their relationship with the president as critical to their effectiveness. The president's evaluation is most important in determining his effectiveness; "the executive director must convince the president that he is running the organization in a manner representative of the president's style."

In daily operations the authority of the executive director is rarely questioned by elected officials; he is paid to keep the society running. However, on particular issues, when those in possession of formal authority are determined to exercise their prerogatives, the influence of the executive may be limited. Nevertheless, he remains the chief policy maker; he tells the council what policy should be.

The executive director's position enables him to raise issues he believes are important and conversely to prevent other issues from being considered. "One way or another he usually generates most proposals for change." At the same time "he can kill an issue just by ignoring it." As head of the managerial staff he may marshal significant resources to develop arguments and gather data in support of or in opposition to a particular issue.[54] Moreover, after a decision has been made, he may continue to exercise influence over implementation. One executive director admits:

Sometimes if I disagree with the officers of the [society], I drag my feet. A staff can really sabotage a program by dragging along on it.... The staff can get things done or it can slow the whole institute down.

The necessity of applying influence through covert channels, by means of suggestion or persuasion rather than overt pressure, makes "salesmanship" a quality crucial to the sucess of executive secretaries. According to one executive secretary:

Staff heads have to have promotional and persuasive abilities. They must propose changes to be made and prepare arguments to support them, all the time aware of the board's veto power. The key to an executive secretary's success is the ability to be a good salesman.

Other suggest a strong resemblance between the role of executive director and the managerial head of large corporations. Thus, one executive director states that a substantial part of his job is "running a business venture." Another suggests that like a corporate manager he "sets policy and develops a corporate point of view"; he then must then act as a salesman to "gain the endorsement of the board."

The perception that specific skills are necessary to carry out their functions effectively leads some executive directors to view themselves as members of a unique occupational catagory; one differentiated from the profession they serve. Executive directors of various societies have even come together in their own associations: MOSS, the Managers of Scientific Societies; and CESSS, the Council of Engineering and Scientific Society Secretaries. These associations provide a forum for discussion of common problems. Moreover, they symbolize a collective consciousness of status. Executive secretaries may or may not possess formal training in the scientific field encompassed by their society. Nevertheless, they perceive their positions as contributing to a distinct career. The executive secretary of one society, a trained scientist, indicates that if he left his present job he would seek another directorship of a scientific society rather than practice science. Thus, it appears that influence in scientific societies has gravitated to individuals whose occupational identifications are not with the profession they serve; their careers are more closely linked to management than to science.

Limitations on Managerial Influence

The influence of managers, substantial as it is, is not unlimited. As indicated earlier, elected officials retain formal authority to govern. In addition, there are other sources of influence in scientific societies which preclude the exercise of monopolistic influence by managerial staff.

Managerial influence is limited in decisions requiring specialized scientific expertise. Even when managers are scientifically trained, specialization limits their ability to participate in some decisions. Indeed, specialization has led to the delegation of policy-making authority within certain defined areas to committees. Societies like the AMS and the AIAA have dozens of committees, many of which operate with virtual autonomy. In some societies there is a tendency for committees to become self-perpetuating; there may be no limit on length of tenure and members may develop proprietary rights to their positions. The executive director of one society suggests that

The committees of [the society] tend to be self-generating and there is great insularity from outside influence. The committees become select clubs and protect their independence.

The proliferation of committees has resulted in a diffusion of authority and has limited the influence of elected officials and managerial staff alike.

Another limitation on the influence of both managers and elected officials is the federal nature of most scientific societies. Societies have hundreds of local and regional sections that carry out many of their services and programs. In varying degrees, depending upon the extent to which a local section possesses resources independent of the national organization, decisions made at the local level may be insulated from the influence of national officials. Local sections elect their own officials and often run their own programs independent of the national society.

Finally, scientific societies are subject to influence from external sources which limits the autonomy of both managers and elected officials. Most societies have corporate memberships; that is, industrial and govermental institutions pay dues and take advantage of publishing and other services of the society. These corporate members serve as a financial infrastructure for some societies. [55] For the most part, the corporate members remain aloof from direct involvement in their daily affairs. Nevertheless, on occasion they may exercise decisive influence. For example, the aerospace industry was primarily responsible for promoting a merger of the American Rocket Society and the Institute of Aerospace Sciences. Moreover, the consolidated institution, the AIAA, remains dependent on industrial support. Indeed, one AIAA leader admits that "it is easy for an institution like the AIAA to be pushed around by industry."

External sources of influence on society affairs are not limited to industry. The Weather Bureau has been a major influence in the AMS. In its formative years an overwhelming majority of its members were in some way connected with the Bureau. As one AMS leader put it, for many years "it appeared that the AMS was little more than an extension of the Bureau." The reluctance of the AMS to become involved in public policy was in part related to the fact that most of its members were civil servants. It was not until the 1950s, with the development of private meteorology, that the Bureau's influence on the society decreased.

The lack of formal authority, the growth of specialization and federalism, and the dependence on external resources all serve to limit the influence of managers. Despite these limitations managers retain significant influence, and scientific societies are best described as limited managerial oligarchies.

Like other voluntary associations, scientific societies manifest the forms but
not the substance of democracy. Thus, there have been few cases of organized
opposition to prevailing oligarchies. Moreover, elections are frequently noncompetitive, and in any case managers as appointed officials are not directly accountable to their memberships. Lack of opposition suggests that members are unper-
turbed by their apparent lack of influence.

It is possible, although we can only speculate in this regard, that most members
accept the fictional characterization of their societies as representative meritocracies: that they see political competition as unnecessary or inappropriate to
their institutions. An alternative explanation is that there have existed few incentives for members to take an active interest in the affairs of their institution. So
long as the functions of societies were limited and their influence on professional
life minimal, participation may not have appeared to be crucial. Consequently,
members may adopt a "very limited relation to the organization"; they may give
"little or no time to the organization" since they are not "guided by its pronouncements."[56]

If the latter interpretation is correct, then we should expect that, as scientific
societies become more influential private governments, the incentives for member
participation will increase. There are sporadic indications that this is happening.
Members of the AIAA have become more active in making demands on that institution as they have been faced with economic difficulties and as they look to the institute to serve in their behalf. In the American Physical Society there has been a
"grass roots movement" relating to the issue of involvement in public policy.[57]
Other societies have also experienced sudden upswings in demands for participation
on particular issues.

Despite such indications of greater activity, on the whole there remains little
member participation in the governance of their societies. Most issues are resolved
at the upper levels of policy making, and most conflict appears limited to confrontation among leaders. One AIAA leader concludes that "member participation occurs
largely on a technical level, and most members remain uninvolved in the politics
and governance of the Institute." Real obstacles to participation aside, there is a
temptation to speculate that oligarchy in scientific societies has arisen, at least in
part, by default of members. Even if incentives increase, greater participation in
internal government may not be forthcoming. In other voluntary associations, and
particularly in other professional associations, member participation often has remained minimal even when there appeared to be much at stake.

COOPERATION AND COMPETITION AMONG PRIVATE GOVERNMENTS

Relations among scientific societies as private government are inconsistent; societies cooperate in sponsorship of meetings, publications, and other services at the same time as they compete for members, income, and prestige. Inconsistency is particularly characteristic of relations among societies representing overlapping fields. In this regard specialization, as it has contributed to the fragmentation of scientific fields, has also contributed to the development of multiple societies with overlapping and competing interests.

Only in the latter part of the nineteenth century did "scientific fields, such as chemistry, physics, astronomy, geology, and biology seem clearly defined by conceptual boundaries."[58] Subsequent specialization resulted in fragmentation; specialities emerged at the edges of a parent body of knowledge or as composites "of knowledge and techniques of several fields."[59] Workers in the new fields were "half-breeds, unwelcome in any established society and forced into separate organizations."[60] The pattern of fragmentation is manifest in the field of chemistry. The America Chemical Society was established in 1876. Specialization continued throughout the late nineteenth century and a "splintering action" began and persisted through the twentieth century.[61] Similar experiences characterize biology, geology, physics, and astronomy. In each case specialization resulted in the establishment of multiple societies all related to a core discipline and engaged in competition for resources.

Competition has occurred when societies in seemingly separate areas broaden their interests and converge on the same area. The American Rocket Society and the Institute of Aeronautical Sciences initially represented different technical areas. After World War II, each society broadened the scope of its activities to include aerospace technology. Other societies, such as the American Society of Mechanical Engineers, also entered the aerospace field and vigorously competed for members and industrial support.

Societies compete for symbolic as well as material resources. Both the American Meteorological Society and the American Geophysical Union (AGU) have interests in meteorology. They cooperate in sponsorship of meetings but compete for international prestige. The AGU, created as an appendage of the National Academy of Sciences, has received the Academy's sanction to serve as the U.S. representative at international meetings. Leaders of the Meteorological Society view this sanction as "an encroachment upon the status of the AMS." They believe that the AMS is the principal representative of American meteorology and that it should be recognized as such at international meetings. The AGU, they propose, has stronger interests in other

areas, such as planetary science. Competition between the societies increased in
the 1950s when the AMS began a concerted effort to "build up its international image."
NAS sanction of the AGU has continued, and it remains an obstacle to the American
Meteorological Society. In addition, a new area of competition emerged in the 1960s.
"As it began to look as though money was going into oceanography, the AMS and the
AGU developed competing interests in the air-water interface." There is little pos-
sibility that competition will cease without some formal affiliation or merger.

   The history of scientific societies in the twentieth century is characterized by
continuous efforts to forestall specialization and, failing that, to affiliate or integrate specialized societies. Frequently attempts to prevent fragmentation take the
form of internal restructuring of existing societies to provide improved services
and status for emerging specialities. For example, the American Association for
the Advancement of Science "preserved practically unaltered its unitary nature"[62]
until the end of the nineteenth century. Since that time the technical work of the
Association has been carried on by affiliated organizations with the AAAS serving
as a unifying body and a recognized spokesman. As specialization proceeded in the
latter nineteenth century, the AAAS responded by dividing into sections representing specialized areas of science

But the specialization movement was too strong for the scientists to be content
merely with forming specialized divisions within the American Association for the
Advancement of Science. The founding of independent specialized societies did not
abate but rather increased, and, "by 1890 the multiplication of special technical
societies had become a recognized potential danger to the Association."[63]

A strategy of co-optation through affiliation began in 1899. By 1910, thirty societies
affiliated; the process continued throughout the century so that the Association now
has about three hundred affiliates.

   Efforts to bring together various societies within a core discipline result in federations and consolidations. In the field of physics, federation occurred with the
establishment of the American Institute of Physics in 1931. The American Physical
Society, created in 1899, served for fifteen years as the only institution in the field
of physics. Subsequently, specialization within physics resulted in a proliferation of
new societies. By the late 1920s, the leaders of the APS became concerned about
fragmentation. The APS appeared to be experiencing trouble in meeting its publication expenses, and members hoped that some form of affiliation would alleviate part
of the burden. Direct consolidation of smaller societies into the APS was not considered feasible since it was expected that groups would fear domination by the APS.
Consequently, in 1931 the AIP was established, consisting of the American Physical

Society, the Optical Society of America, the Acoustical Society of America, and the Society of Rheology. Each society retained control over the technical aspects of its work and looked to the AIP for administration of programs and services.

In addition to federations, societies came together through direct merger or consolidation into a single institution. The establishment of the American Institute of Aeronautics and Astronautics resulted from consolidation of the Institute of the Aerospace Sciences and the American Rocket Society in 1963. Prior to World War II, the areas of interest of the IAS and the ARS were clearly separate. The IAS was an aeronautical society that disclaimed interest in rocketry; the ARS was a propulsion society with little interest in aeronautical applications. After World War II, the interests of the two societies resulted in joint memberships and similarities in the subject matter of meetings and publications; convergence of interests fostered duplication of services and competition for resources. The logic of merger was predicated upon the assumption that, as duplication led to inefficient utilization of resources, competition resulted in a smaller share of available resources for each institution. Consolidation was offered as a means of eliminating competition and waste and providing a single organization representative of the profession as a whole.

Despite continuous efforts at federation and consolidation, specialization continues to reinforce the fragmentation of scientific societies. No institution has emerged that can operate as an influential private government in the scientific community as a whole. The private government of professional science remains pluralistic.

PRIVATE GOVERNMENT AND PUBLIC GOVERNMENT

As scientific societies have become more influential private governments, they have also become more deeply involved in public government. Contemporary societies are significantly more active in public affairs than their nineteenth and early twentieth century counterparts. At the same time such activities frequently have been engaged in with reluctance. There has been and continues to be intensive debate within these societies concerning their public role.

The controversy is deeply embedded in the history of science. It was particularly acute in the nineteenth century as the scientific community reevaluated the earlier emphasis on the social utility of science and as commitment to the pure science ideal developed.[64] Nevertheless, scientific societies traditionally remained aloof from political involvement, often appearing unconcerned with "the world beyond their particular disciplines."[65] Consequently, while the present controversy has deep historical roots, "only in the past decade or so have the learned societies themselves edged into the realm of public affairs."[66]

353.0085 T233s

c.1   25

Lobbying

Scientific societies have not engaged in overt lobbying. Their reluctance to participate in direct political action is not a result of a lack of self-interested motivation.[67] Rather, some members remain committed to a belief that if science is to
flourish it must maintain independence from political involvement. Embodied in
such a perspective is the view that the search for scientific truth is incompatible
with the search for power. It is feared that insofar as scientists become engaged
in the search for power they become subject to political responsibilities that necessarily limit their freedom in pursuit of truth.[68]

A more pragmatic source of resistance to political involvement is the belief that
short-term gains to be derived through lobbying are weighed by losses of public
credibility and status. It is argued that the public status of science is under attack
from "a new kind of rebellion" that challenges the conviction that scientific development is in the vanguard of political and economic progress.[69] Lobbying would reinforce perceptions that scientists and scientific institutions are basically self-
serving. Consequently, it is proposed that scientific societies should seek to gain
public support by providing "objective analysis of major problems in public policy
that fall within their technical competence."[70] Thus, a past president of the AIAA
states that societies "should speak out in support of research and development, but
... should avoid espousing specific budgetary causes that might appear to be motivated by self-interest...."

Some members of scientific societies object to lobbying because they believe that
taking political stands would create internal dissension detrimental to the purposes
of these institutions. This view is reinforced by professional ethics of noncompetition and beliefs in the necessity of presenting a united front vis-a-vis society. It
is believed that attempts to achieve consensus on public issues would result in divisiveness that might threaten the public status of the profession.[71] Moreover,
given the oligarchic tendencies of scientific societies, attempts to lobby without
the development of consensus are certain to raise questions of legitimacy.[72]

Some leaders suggest that scientific societies lack the resources to be effective
lobbying agents. An AIAA leader argues that "societies are really pretty weak institutions; leaders are better off exercising political influence informally behind
the scenes than engaging in concerted public efforts." Since lobbying is expensive,
it may divert funds from other functions and services. In some societies rising
costs of publications and meetings have led to increases in member dues. Leaders
of these societies may then question the willingness of the members to bear the
additional financial burdens imposed by lobbying.

The most frequently perceived economic obstacle to political involvement is possible loss of tax-exempt status. Section 501 (c) (3) of the Internal Revenue Code exempts societies from federal taxes and allows members to deduct payments of dues from their own taxes. The code provides that "no substantial part" of the activities of an organization in this category may be devoted to "carrying on propaganda or otherwise attempting to influence legislation." It also prohibits support of candidates for public office.

Leaders of societies have found the code vague in its formulation. Consequently, as former Congressman Emilio Daddario points out, "some organizations have taken an extreme position and read 'no substantial part' to be, in effect, 'no part.'"[73] In the past the Internal Revenue Service has added to the confusion by stating that what constitutes a 'substantial part' must be evaluated on the total merits of each situation.[74]

If a society loses its status under 501 (c) (3), it still may qualify for tax exemption under other sections of the code. However, the economic effects are substantial. Despite continued exemption of the society from federal tax, members would lose their personal tax deduction. Consequently, the attraction of new members becomes more difficult. In addition, societies may also lose exemption from state and local taxes. Finally, there is reason to believe that maintenance of status as a "charitable" institution is itself a political resource. K. M. Reese points out that the "charitable" organization acquires almost automatically a mantle of credibility and an image of political neutrality.[75]

Some efforts have been made to amend the Internal Revenue Code. Senator John Sherman Cooper of Kentucky introduced an amendment to the Tax Reform Act of 1969 to make possible a more active public role for societies; specifically, to allow them to provide Congress and government agencies with information and advice.[76] The amendment was not enacted. However, members of Congress and the executive branch continue to believe that the participation of scientific societies in the policy-making process is important. Former Congressman Daddario has persistently argued that Congress needs these societies to present their opinions and analyses of public issues. He believes that, insofar as these societies remain aloof from political involvement, the legislative process suffers.[77] Richard Carpenter has reached much the same conclusion: "professional technical societies can be a most valuable source of advice and information to Congress."[78] Moreover, "a standing invitation to communicate with the Congress has been issued repeatedly by members of both Houses ... ."[79]

IRS spokesmen recently suggested that the status of the societies need not be endangered if the initiative for the involvement in public policy is taken by Congress and if societies do not devote substantial amounts of resources to advocating a particular legislative position. On this basis some societies have made their views on certain policy issues known to both the executive and legislative branches. The Ecological Society of America, for example, has provided witnesses before congressional committees concerned with ecological issues and, with the support of the National Science Foundation, has promoted the establishment of a national institute of ecology. Such examples notwithstanding, in almost every case scientific societies have steered clear of lobbying on particular bills. Lobbying has remained the province of splinter groups and organizations usually created expressly for that purpose.

The Nature of Political Involvement

Since 1962 the National Academy of Sciences' Committee on Science and Public Policy has supplied Congress and the executive branch with information and advice based on studies of subjects ranging from technology assessment to status reports on various disciplines. The more specialized societies have followed the lead of the NAS in creating similar committees that perform the symbolic function of affirming the legitimacy of some type of involvement in public affairs.

Some of the committees have responsibility for initiating and carrying out studies on public policy and for publicizing and disseminating the results of such studies. Thus, the American Chemical Society's Committee on Chemistry and Public Affairs recently issued a report on chemistry and the environment which was undertaken in response to a 1965 study on environmental quality sponsored by the President's Science Advisory Committee. The ACS committee believed that the earlier study inadequately considered aspects of environmental issues relating to chemistry. The ACS study is purportedly objective; yet leaders of the ACS are well aware of the potential for charges of bias. As Harvey Brooks has pointed out, an objective study of chemistry by chemists is "ipso facto suspect."[80] Government leaders also recognize the difficulty facing societies conducting technical study of policy problems:

It is difficult for such a group to issue a report which will not be judged as biased in some way or another. If the document is critical of Federal agency practices, it must have overwhelming compilation of evidence. The work is done largely by persons in universities and industry on a part-time basis. These reasons explain the fact that such studies are rather slow in materializing. In the meantime, agency programs gather momentum so that the eventual report takes on the character of ex post facto review and recrimination rather than being useful as an aid in planning Government actions. [81]

The issue of bias is complicated by economic factors. In many cases the studies conducted by scientific societies are paid for by agencies of the government which, like the societies themselves, might benefit from the recommendations of study groups. Thus, K. M. Reese points out that

Some will regard as incestuous, for example, a situation in which NSF supports studies by NAS that recommend increased support for research when much of the support would be supplied by NSF itself. [82]

Reese suggests that the same dilemma is evident in the American Chemical Society's financial contributions to a National Academy of Sciences study recommending increased federal support for chemistry. [83] The study may be objective; nevertheless, the American Chemical Society has a vested interest in the outcome. It is unlikely that an institution would spend thousands of dollars on a project which was expected to reflect negatively on the profession it serves.

Charles Price, founder and chairman of the ACS's Committee on Chemistry and Public Affairs, suggests that scientific societies limit their public activities to "issues in which a fairly significant component of science know-how is involved."[84] Scientists ought concern themselves only with issues to which they can bring scientific and technical expertise. Nonetheless, it is difficult to determine which issues are of a technical nature and to separate their technical and nontechnical aspects. Various societies have interpreted as falling within their domain of interest: cutbacks in federal funding, selective service requirements, use of herbicides in Vietnam, and weather modification. There is no precedent. What one society believes is an appropriate concern, another finds beyond the limits of its legitimate interest and purpose. The American Meteorological Society refused to endorse a public statement of its Committee on Atmospheric Problems of Aerospace Vehicles. Members of the committee, who were also members of the AIAA, took it to the institute and received endorsement from that organization. Ironically, despite its reluctance in this case, the AMS was one of the first institutions to become involved in public affairs;[85] the AIAA has only more recently become active.

At times public affairs committees have been created without an immediate purpose but in anticipation of a future need. The AMS recently created a Committee on Public Policy to be formally responsible for issuing public policy statements. The Committee was created, in part, to provide the AMS with a mechanism that could respond to sudden government actions affecting meteorology.

Some societies have established committees so as to indicate interest in public affairs without becoming involved in more direct political action. In the last few years the American Physical Society has had exceptional difficulty in determining the nature of the public affairs role it should adopt. One APS leader indicated that the question facing the society was not "whether to become involved in politics but rather in what ways to become involved." Some members of the society have attempted to amend the APS constitution so as to enable the APS to take positions on public issues such as the Vietnam War. The amendment was defeated, and some dissident members formed a separate institution, the Scientists and Engineers for Social and Political Action. The new institution was to seek new and radical solutions for long-range and immediate problems and to press for effective political action. At the same time, the American Physical Society responded to a petition of some of its members by urging that a new division of the society be devoted to the study and discussion of physics and society. The society eventually created a Committee on Public Policy, an indication that it has some continuing interest in public affairs. Similarly, the American Institute of Physics, of which the American Physical Society is the largest member, has thus far resisted efforts by some physicists to make the institute active in lobbying for the economic interests of the profession. However, the institute has created a Committee on Physics and Society whose initial efforts included studying physics manpower and changes in federal funding of physics.

In addition to conducting technical studies on public policy issues, scientific societies have sponsored forums and symposiums on science and public affairs. The AIAA has sponsored public forums ranging in subject matter from the future of civil aviation to science, technology, and the quality of life. Many societies also publish political news and opinion affecting the profession in their journals and newsletters. These activities reinforce collective consciousness of status within the profession by providing members with information concerning relations of the profession to public government and fostering common views of self-interest. Generally, public affairs activities contribute to the political socialization of members.

Corrine Gilb has suggested that professional associations in fields such as medicine and law operate as preliminary arenas of public lawmaking; that is, they essentially preempt the authority of public government to make law which affects the profession.[86] Thus far, scientific societies have not been active in this way. We should expect however that, as they seek greater professional autonomy through licensing and enforcement of ethical standards, scientific societies will also attempt to preempt the authority of public government. In addition, scientific societies have already become active as preliminary arenas of public lawmaking insofar as public law is becoming increasingly dependent upon expertise. Some government officials foster the participation of scientific societies in public affairs so that they may mobilize professional skill in the formulation of public law.

As they have become more influential private governments, scientific societies also have become active in regulating the relationship between their professions and public government. Yet such activity has been pursued with reluctance, and scientific societies have generally remained aloof from direct forms of lobbying. At the same time it would be difficult to interpret their public affairs activities as other than self-interested. Like other political groups, they expend energy and resources in the hopes of creating a more generally favorable climate of opinion. As government has emerged as the principle patron of research and development, it becomes important to maintain and nourish a broad base of public support for such patronage. The public affairs activities of scientific societies thus far have not been effective in mobilizing support. However, leaders of these societies as well as government officials expect that these private governments will adopt a more active role in public policy formulation.

CONCLUSION

Scientific societies are becoming more influential private governments; they are exercising control over the professional lives of their members, and they are regulating relationships between their professions and society. Nevertheless, in comparison with professional associations in other fields, scientific societies are weak private governments; they have not manifested great effectiveness as vehicles of professional autonomy. Moreover, the operation of scientific societies as private governments poses serious questions of legitimacy. Specifically, who or what is the legitimate constituency of scientific societies? Like other contemporary private governments, scientific societies have encountered difficulty in identifying purposes and responsibilities.

To some extent the dilemmas facing contemporary societies are symptomatic of the confusion and uncertainty that have characterized the scientific community as a whole since World War II. The rapid and growing interdependence of science and society has resulted in much bewilderment concerning how science and society ought to be related. Old standards no longer seem to apply, but new standards have not emerged.

While the variety of views concerning the legitimate constituency of scientific societies appears virtually endless, there are major themes that persist in debates on such issues. One view finds its point of reference in the late nineteenth century with the emergence of what George Daniels has called the "pure science ideal."[87] It embodies a belief that science ought to be pursued for its own sake, insulated from social responsibilities and social influence. Scientific work is to be differentiated from professional work in that the former is concerned only with the advancement and diffusion of knowledge and not with its application. Scientific societies then ought to limit their activities to services such as the publication of journals and the sponsorship of meetings which foster scientific development by contributing to interaction.

An alternative view suggests that scientific societies are, or at least ought to be, like professional associations in fields such as medicine and law. The principal constituency of these associations is the professional community. Serving that constituency necessitates concern for professional practitioners not only in fostering intellectual development but also in terms of social status and economic welfare. Scientific societies then are responsible for promoting the interests of members and for exercising control over the direction of the profession as a whole. Moreover, insofar as scientific development is crucial to society, then the welfare of the profession contributes to social welfare. Acting on this basis, scientific societies ought to become influential private governments seeking autonomous control over the professional lives of members and over the relationship of the profession to society.

Finally, there is a view of science as a public resource. Embodied in this perspective is the belief that society is the principal constituency of science. Such a view has points of reference very early in the history of modern science; it is basic to the Baconian notion of scientific institutions dedicated to the promotion of knowledge which may improve the condition of man. It is also a view with more contemporary roots: conceptions of scientific manpower and institutions as a social resource deserving public patronage insofar as they contribute to social welfare. Indeed, this view is implicit in the current legal characterization of scientific societies as "charitable" institutions. At the same time, with modification, this theme is consistent with some elements of the "current rebellion" against science. If scientific development is detrimental to public welfare, then it should lose public support and perhaps even become subject to public restriction. From this perspective scientists and scientific institutions have above all else a definite social responsibility.

Contemporary scientific societies have been faced with the challenge of achieving some accommodation among these views. The question remains, however, as to what extent such views are compatible. Moreover, each view embodies different potential costs and benefits; each leads to a different balance between the values of autonomy and social responsibility. It is unlikely that scientific societies will reach a sudden consensus; it is more likely that decisions will continue to be made incrementally and that cumulatively they will emphasize particular responsibilities. Incremental decisions have led scientific societies to evolve complex, and often seemingly inconsistent, functions. It should not be expected then that the dilemmas of legitimacy will be resolved rapidly by incremental steps. Moreover, to some extent the directions that scientific societies adopt will be a function not only of their own choice but also of the constraints imposed by society, particularly public government. It is doubtful, for example, that the scientific community and its institutions can retrench completely to a committment to the pure science ideal when government is the principle source of support for research and development. In any case, the dilemmas of legitimacy facing contemporary scientific societies cannot be resolved unless a more definite identification of standards for relating science to society is established. Moreover, the operation of scientific societies as private governments will continue to engender problems so long as legitimacy is uncertain.

NOTES

1. Charles Merriam, Public and Private Government (New Haven: Yale University Press, 1944), p. 6.

2. Our definition includes technical but not engineering societies. The word technical, rather than engineering, is used advisedly. We have not studied purely engineering societies; that is, societies composed almost exclusively of engineers and oriented toward the advancement of an engineering field, for example, The American Society of Mechanical Engineers. These differ from technical societies in that the latter are hybrids; that is, their memberships tend to include scientists and engineers from various specialized fields all concerned with aspects of a particular technology such as aeronautical or marine technology. In the body of this paper, we use the term scientific society to refer to both scientific and technical societies.

3. Case studies were conducted of the American Institute of Aeronautics and Astronautics (AIAA), the American Meteorological Society (AMS), and the American Institute of Physics (AIP). In addition, data on other societies were collected from journals and other published material distributed by such institutions.

4. Corinne L. Gilb, Hidden Hierarchies (New York: Harper & Row, 1966), p. 109. Gilb's book includes the most comprehensive treatment of professional associations as private governments.

5. Ibid.

6. See William J. Goode, "Encroachment, Charlatanism, and the Emerging Profession," American Sociological Review, Vol. 25, December 1960, pp. 102-114.

7. The elements of professionalization presented in this section represent an adaptation of a framework used by George Daniels in his study of scientific professionalization in the nineteenth century. See George Daniels, "The Process of Professionalization in American Science: The Emergent Period, 1820-1860," ISIS, Vol. 58, Part 2, 1967, pp. 151-166.

8. Ibid., p. 152.

9. William Goode, "Community Within a Community: The Professions," American Sociological Review, Vol. 22, April 1957, p. 196.

10. Theodore Caplow and Reece J. McGee, The Academic Marketplace (Garden City, N.Y.: Doubleday & Co., Inc., 1965).

11. William Kornhauser, Scientists in Industry: Conflict and Accommodation (Berkeley, Calif.: University of California Press, 1962), p. 87.

12. Gilb, Hidden Hierarchies, p. 69.

13. Kornhauser, Scientists in Industry, p. 87.

14. Such a plan is now under consideration by the American Chemical Society. In effect it would reduce long-term financial dependence of the individual member on a particular job and increase his dependence on participation in a society.

15. The APA argues that none of the existing societies has accepted a mandate "to concern themselves with the economic and professional well being of the people in the profession," and moreover, "they have no significant funds to channel towards solution of the economic problems currently facing physicists." ("Program of Action," American Physicists Association, Washington, D.C., 1970).

16. Ibid.

17. Kornhauser, Scientists in Industry, p. 89.

18. Anselm Strauss and Lee Rainwater, The Professional Scientists: A Study of American Chemists (Chicago: Aldine Publishing Co., 1962), Chapter 10.

19. Kornhauser, Scientists in Industry, p. 90.

20. Ibid., pp. 103-114.

21. Interview with David Ludlum, American Meteorological Society Staff, Boston, Mass., July 1969.

22. Gilb, Hidden Hierarchies, p. 67.

23. Seymour Martin Lipset, Martin A. Trow, and James S. Coleman, Union Democracy (Garden City, N.Y.: Doubleday & Co., Inc., 1962), pp. 270-273.

24. Gilb, Hidden Hierarchies, p. 67.

25. A different objection to the establishment of a code has been that it is inconsistent with the nature of scientific pursuits; that is, ethical standards are believed

implicit in scientific methods of research. In technical societies like the AIAA, engineering members appear more favorably disposed to a code than the scientific members (James Harford, executive director of the AIAA, personal communication).

26. A recent unpublished survey conducted by Anatol Rappaport under the auspices of the AAAS indicates broad support among scientists for the establishment of formal ethical standards. Study cited in Bulletin of the Atomic Scientists, Vol. 20, September 1968, pp. 39-40.

27. E. C. Hughes, quoted in Gilb, Hidden Hierarchies, p. 62.

28. Ibid.

29. Interview with James Harford, executive director of the AIAA, New York City, July 1969.

30. Interview with John Newbauer, editor of Astronautics and Aeronautics, AIAA, New York City, July 1969.

31. Interview with Alfred G. Kildow, director of public affairs, AIAA, Anaheim, Calif., September 1969.

32. Strauss and Rainwater, The Professional Scientist, p. 70.

33. Interview with Alfred G. Kildow, AIAA, September 1969.

34. Interview with Eugene Kone, director of public relations, AIP, New York City, March 1970.

35. Ibid.

36. Ibid.

37. Gilb, Hidden Hierarchies, p. 70.

38. Ibid.

39. Interview with Eugene Kone, AIP, March 1970.

40. Philip Selznick, The Organizational Weapon (Glencoe, Ill.: Free Press, 1960), p. 96.

41. Lipset, Trow, and Coleman, Union Democracy, pp. 3-4.

42. "Constitution of the American Meteorological Society," reprinted in The Bulletin of the American Meteorological Society, Vol. 45, August 1964, p. 17.

43. Ibid., p. 19.

44. Gilb, Hidden Hierarchies, p. 113.

45. Ibid.

46. Ibid.

47. Some societies like the AMS nominate at least two candicates for each office. However, others like the AIAA nominate only one candidate for some offices, including president. There is no case in which a candidate offered by the AIAA nominating committee has been challenged by nominations through petition. This may reflect lack of opposition, but it may also reflect a difficulty in fulfilling petition requirements. Officers of the AIAA have defended noncompetitive elections, arguing that competition for office would be detrimental to the professional interests of the Institute.

48. Thus, since its inception, the AIAA has not had more than 35% of its ballots returned. This may reflect the lack of competitive elections in the AIAA; that is, the lack of choice may undercut a potential incentive to participate in the electoral process.

49. In the remainder of this chapter quotations from interviews conducted on non-attributable bases will be presented without footnotes.

50. Thus, for example, until 1946 the AMS had no paid staff. At the end of the war the society underwent a program of reorganization of services and structure. The administrative pressure of expanded services led to the hiring of an executive director and subsequently a staff to supplement his efforts. Similarly, the ARS appointed its first executive secretary in 1953 at a time when it was attempting to solidify its status as a professional propulsion society; his principal task was to coordinate the society's program of expansion.

51. Thus, the AMS had one executive director serving for twenty-five years; the IAS had two over a thirty-two-year period; and the ARS has one who served ten years and then became executive director of the consolidated AIAA.

52. Gilb, Hidden Hierarchies, p. 132.

53. Ibid.

54. Thus, one executive director suggests that his influence is particularly great in initiating and challenging policy. "I can get the president and board to consider things I think are important and I can challenge decisions they make. ..."

55. In addition to corporate dues, support is manifest in advertisements in publications, exhibits at meetings, and contributions to employee participation.

56. Selznick, The Organizational Weapon, p. 96.

57. Some members organized a movement to amend the society's constitution and thus make possible taking stands on issues such as Vietnam. The amendment was defeated, and the dissident members formed their own society. However, the movement seems to have had some long-term success in that the American Physical Society is formally reevaluating its political role.

58. David Van Tassell and Michael G. Hall, Science and Society in the United States (Homewood, Ill.: Dorsey Press, 1966), p. 33.

59. Ibid.

60. Ibid. For example, Van Tassell and Hall point out that "Mendel's theory of inherited characteristics was not lost, but resisted, partially because he applied

statistics to botany; and in 1901 Karl Pearson and Francis Galton founded the journal Biometrika in response to the Royal Society's decree that mathematics should be kept apart from biological applications." pp. 33-34.

61. Ralph S. Bates, Scientific Societies in the United States, 3rd ed. (Cambridge, Mass.: The MIT Press, 1965), Chapter III. For example, the American Leather Chemists Association was established in 1903; the American Institute of Chemical Engineers in 1908; the American Oil Chemists Association in 1909; the American Institute of Chemists in 1923; the Association of Consulting Chemists and Chemical Engineers in 1928; the Wood Chemical Institute in 1929; the Metropolitic Micro-chemical Society in 1936; and the American Society of Brewing Chemists in 1945.

62. Ibid., p. 125.

63. Ibid.

64. Thus, Arnold Thackray points out that the current debate should not be viewed as a sudden novelty. Arnold Thackray, "Reflections on the Decline of Science in America and on Some of its Causes," Science, Vol. 173, July 2, 1971, p. 28.

65. K. M. Reese, "Scientific Societies and Public Affairs," Chemical and Engineering News, Vol. 48, May 3, 1971, p. 30.

66. Ibid.

67. K. M. Reese argues that self-interest "is the driving force behind the scientific and engineering societies efforts in public affairs." Thus most societies became involved in public affairs since 1965 concurrent with a leveling off in federal support of research and development. Ibid.

68. Don K. Price suggests that: "The dilemma is ages old—the dilemma between truth and power, or, rather, between starving in the pursuit of truth and compromising truth to gain material support." Don K. Price, "Purists and Politicians," Science, Vol. 163, January 3, 1969, p. 25.

69. Price elaborates: "The ideology of the rebellion is confused; you can find in it little clarity or consistency of purpose....From the point of view of scientists, the most important theme in the rebellion is its hatred of what it sees as an impersonal technological society that dominates the individual and reduces his sense of freedom. In this complex system, science and technology, far from being considered beneficent instruments of progress, are identified as the intellectual processes that are at the roots of the blind forces of oppression." Ibid.

70. Reese, "Scientific Societies and Public Affairs," p. 30.

71. Government officials, when they have advocated a more active role for scientific societies in public policy, have stressed that such participation does not require that the societies reach consensus on public issues. For example, Richard Carpenter proposes: "The technical societies are not expected to present a consensus view on issues—which might engender internal strife. Rather, they should provide a forum for discussion of science policy issues.... The policy discussions, with the participation of society leaders, would provide interpretations and viewpoints of great usefulness to the Congress." Richard A. Carpenter, "Science, Policy and Congress," Midwest Research Institute Quarterly, Winter 1968-1969, p. 5.

72. In this respect, however, scientific societies would be similar to most other interest groups. In most interest groups the controlling oligarchy tends to play a determining role in the formulation and presentation of political views. Moreover, political scientists have suggested that the oligarchic nature of organization helped strengthen interest groups. At the same time oligarchy has been a major focus of critics of pluralist theories. See H. R. Mahood, "Pressure Groups: A Threat to Democracy?," in H. R. Mahood, ed. Pressure Groups in American Politics (New York: Charles Scribner's, 1967), pp. 296-298.

73. Emilio Daddario, "Scientists and Legislators," BioScience, Vol. 19, February 1969, p. 149.

74. Present criteria tend to inhibit but not completely prohibit political activity. In one court case it was found that devoting less than 5% of resources to political activities did not consitutute a substantial part of overall functions. In a more recent case the IRS revoked the tax-exempt status of the Sierra Club, ruling that more than 50% of the issues considered by the club's board of directors related to pending legislation.

75. Reese, "Scientific Societies and Public Affairs," p. 31.

76. Under the auspices of the American National Standards Institute, leaders of a number of societies, including the AIP and the AIAA, have met together to consider and endorse the amendment.

77. Daddario, "Scientists and Legislators," pp. 148-151.

78. Carpenter, "Science, Policy and Congress," p. 7.

79. Ibid.

80. Brooks, quoted in Reese, "Scientific Societies and Public Affairs," p. 33.

81. Congressional Hearings, House Committee on Science and Astronautics, Sub-Committee on Science, Research and Development, October 1968, p. 44.

82. Reese, "Scientific Societies and Public Affairs," p. 33.

83. Charges of bias are also to be expected when, for example, the American Chemical Society sponsors a study of the effects of chemistry on the American economy and when one leader of the society suggests that its purpose is to identify "what chemistry has done for mankind."

84. Charles Price, quoted in Carpenter, "Science, Policy and Congress," p. 26.

85. Since 1957 the American Meteorological Society has issued policy statements on such matters as weather modification, long-range forecasting, and weather satellites.

86. Gilb, Hidden Hierarchies, p. 109.

87. George H. Daniels, "The Pure-Science Ideal and Democratic Culture," Science, Vol. 156, June 30, 1967.

Eugene B. Skolnikoff

Immediately after World War II, important segments of the U.S. scientific community became embroiled in a public political controversy over the proposed machinery for the control of atomic energy. It was an unprecedented experience for most of the scientists involved, since their political involvement during and and prior to the war had been primarily out of the public limelight. Their success as lobbyists in forcing the substitution of the MacMahon Act for the May-Johnson bill appeared to be a harbinger of a new era in which the scientific community would become a major political force on the American scene.

It did not happen that way. Whatever influence scientists developed in political affairs continued to be exercised primarily within the executive branch. As a political action group, scientists, with a few notable exceptions, refused to organize to exert any consistent pressure and only rarely became involved in anything approaching group action on specific issues. Even their victories over issues such as the MacMahon Act and the test ban treaty proved to be equivocal; the Atomic Energy Commission in practice came to be largely dominated by military requirements whatever the organizational structure, and the atmospheric test ban tended ultimately to accelerate bomb tests rather than impede them.

By the late 1960s the situation apparently began to change, as many of the leading "establishment" scientists of the postwar decades began to move toward increasing public criticism of federal policy for science and technology, and in particular the pattern for use of technology. There were many divergent reasons that led to this development, including prominently the excesses of Vietnam and the seemingly obvious mismatch in the allocation of resources for science and technology in light of the growing social problems within the United States. What actually led particular scientists to become involved publicly, as Anne Cahn shows so well in her study, was a wide variety of individual views and experiences, often triggered by a sudden personal involvement such as a plan for a missile emplacement in a scientist's hometown.

In a sense, we are seeing now a major new attempt to mobilize the scientific community as a public critic of policy, an attempt with greater apparent political depth than was demonstrated in the intervening years in, for example, the civil defense or test ban debates. Even the organization established after the war as the spearhead of political action—the Federation of American Scientists—is undergoing an astonishing rebirth of energy and ambition.

The debate over the ABM is the major benchmark in this new development. It provided the most direct and public confrontation between elements within the scientific community, and between prominent members of the community and the executive branch of government, since the atomic energy debate of 1946. In parallel with that earlier debate, it was heralded as a symbol of a new willingness of the scientific community to stand up and be counted on the political issues growing out of their work in the laboratory and in industry.

It is much too early to judge whether the parallel will continue and scientists will once again withdraw from public controversy and scrutiny. There are signs pointing both ways. The judgment can be made, however, that, whatever the scientific "community" does, a substantial segment has tasted public action, has been well received by public figures—especially in the Congress—and shares a conviction that the potential for technological mischief is so great that it has a responsibility to act.

It is of great interest and importance, therefore, for students of politics to understand what this changed role of at least some scientists has meant and what it is likely to mean in the future. One of the obvious places to start is the ABM debate itself, and this is what Anne Cahn has done with insight and imagination. Why scientists became involved, to what degree and for what reason, with what common goals, and to what effect (as perceived by the scientists) are all relevant issues in the study.

The results of the work cannot be summed briefly; I leave the exposition to be done in her own words. But the results are fascinating, and often surprising, in their indications of the haphazard nature of the stimuli, the strange communications links among the scientists, and (less surprising) the importance of career patterns in determining the positions individuals took in the debate.

What clues the study provides for those interested in stimulating more political action by scientists are best left to the idiosyncrasies of the readers. To one reader at least, who believes scientists and engineers must take a more prominent public role in policies regarding technological development, the conclusions of the study are often disheartening, and do point in effect to the need for self-conscious programs in the private sector designed to provide a framework for analysis and political action on technologically related issues. If this is so, it is essential to develop such programs today; the need is great.

Anne Hessing Cahn

PROLOGUE

On September 18, 1967, Secretary of Defense Robert S. McNamara announced
that the United States would deploy an antiballistic missile system (ABM). Almost
two years later, in a crucial vote on August 7, 1969, fifty United States Senators
voted against an ABM deployment.

In the twenty-three months between these two events, the American scientific
community became galvanized, energized, and polarized in attempts to reverse
or expand the former decision and to affect the latter one. Hundreds of members
of what has been termed "an apolitical elite" testified before congressional committees, briefed senators and congressmen, and lobbied with them as well; signed
petitions, ran ads in newspapers; wrote articles and reports; edited and contributed
to books, marched to the White House; set up information booths at shopping centers;
traveled across the country speaking to audiences ranging up to several thousand
persons; debated with generals and colonels, senators, and fellow scientists;
addressed state and local officials; worked with women's groups, peace groups,
labor, religious, and civic organizations; presented resolutions to village, town,
city, county, and state legislative bodies; printed and distributed buttons and
bumper stickers; appeared on local radio talk shows and on national television
programs.

As the youngest child asks at the Jewish Passover dinner, "Why is this night
different from all others?" so might one well ask why this arousal, this concern
and this animation about this particular weapon system? Dozens of weapons
systems— large and small, offensive and defensive—have been invented, refined,
and their deployment encouraged by scientists since the end of World War II with
only faint whispers of doubt or hesitation about their desirability or necessity.
Proponents and opponents alike have proclaimed that the ABM became a symbol.
One leading advocate of deployment, Albert Wohlstetter, referred to the debate as
"Good Guys, Bad Guys, and the ABM."[1] Accepting that, indeed, for many scientists the ABM was a symbolic issue through which they could voice their despair
or indignation with the war in Vietnam, with the "military-industrial complex,"
with the specter (or reality) of their own impending unemployment or obsolescence—
it remains a fact that the ABM debate mobilized and "turned on" a sizable segment
of the scientific community. As Alice Kimball Smith, in her chronicle of an earlier

effort of scientists to influence public policy wrote, "A certain epic quality attaches
to any experience by which men are so stirred as to be wrenched from their accustomed patterns of behavior."[2]

Since the advent of the atomic era, the "peril and the hope" of Alice Smith's
book—the interface between scientists and the political world in which they live—
has become a matter of increasing concern to academicians and the public alike.
A voluminous literature has emerged describing such phenomena as The Two
Cultures,[3] The New Priesthood,[4] The Scientific Estate,[5] "The Scientific Adviser,"[6]
and The New Brahmins.[7]

However insightful these pioneering studies were in examining the virgin terrain
of the relationship between scientists and their political environs, the great bulk
of them painted the landscape with broad descriptive strokes. As the field of science
and public policy advances from its infancy, its empirical research must shift
increasingly from the general to the more specific. In order to understand why the
ABM became such a major controversy among scientists in the late 1960s, it would
seem useful to examine the participants in the issue in greater detail.

This study addresses itself to the following questions: Who were the scientists who
participated in the ABM issue? What were their motivations, their expectations of
success, their perceptions of the debate itself and its effect upon them individually
and upon the scientific community? What seems to differentiate the pro- from the
anti-ABM scientist? Are these activated scientists likely to remain as vocal players
at the tumultuous game of politics or will they return to the reputed tranquillity and
quietude of their research? Lastly, what was the effect of all this activity upon the
ABM issue itself? How efficacious do the participants view themselves and how
effective do the decision makers, both within the executive departments and in the
Senate, believe the scientists were?

The Setting
Several social scientists[8, 9] have referred to the elusiveness of a precise definition
of the term "scientist." The Census Bureau's criterion of "trained as" is just as
imprecise and subject to obvious counterexamples as the Labor Department's standard of "employed as." For a study dealing with the political arena, the term must
be defined in the broadest sense possible. Thus, we have chosen to treat as a
"scientist" anyone who considers himself and/or is considered by others to be a
scientist. This conforms to the criterion used to compile the National Register of
Scientific and Technical Personnel, which is based on information supplied by

scientific societies. Broadly construed in this way, the definitional net includes
engineers such as Jerome Wiesner and administrators like James Killian, both
of whom served as presidential science advisors. *

Our definition of participation is based upon Lester W. Milbrath's work. He
has ranked political acts in a hierarchy of political involvement from "spectator"
activities, such as voting or reading campaign literature, through "transitional"
activities like making a financial contribution to a party or candidate or attending
a political meeting, to "gladiatorial" activities that include contributing time in a
political campaign, becoming an active member in a political party, soliciting
political funds, and running for or holding public or party office. [10]

Adapting this terminology to the present case, "gladiators" in the ABM arena
were those scientists who testified, lobbied, wrote and debated or lectured on the
subject of the ABM. However, these overt and public actions represent only one
facet of the scientists' involvement in the ABM issue. By turning the kaleidoscope
just a few degrees, there quickly tumble into view the activities of "transitional"
scientists who served on task forces and advisory panels. The boundaries between
the "gladiators" and "transitionals" are not always sharp and distinct and, indeed,
sometimes overlap. Scientists who acted only as "spectators" and did nothing beyond
signing their name to a petition, buying an ABM bumper sticker or button, joining
in a physicists' march, or voting in the American Physical Society poll were excluded.

A concerted effort was made to compile as comprehensive a list as possible of
scientist-participants. This was initially obtained by a thorough search of the relevant congressional hearings and national, regional, and local newspapers. The
"snowball" method of asking each person interviewed to name all other participants
he knew was also utilized. The effort resulted in a population of 201, of whom 152
(76%) constituted the sample. **

The bulk of the data was gathered by means of personal interviews conducted by
the author between June 1969 and April 1972 (n = 122). The remainder was obtained
from questionnaires. The interviews lasted from just under one hour to three hours

------

* In the course of the research it became apparent to the investigator that one man's
"strategist" may be another man's "social scientist," with the former term often
denoting approbation and the latter disparagement. Regardless of their connotations,
both "strategists" and "social scientists" are encompassed under the rubric of
"scientist" in this study.

**Outright refusals to cooperate in the study were received from only ten scientists
who consisted entirely of contractor and government employees and official government advisers. Based on the comparisons that could be made readily, the nonre-
spondents did not differ in any significant way from the respondents.

or more, with a modal length of one and one-half hours. No tape recorder was used, at least by the investigator, but extensive notes were taken during and immediately after each meeting.

The interviews consisted of open-ended questions focused on the scientist's role in the ABM issue, his perceptions of the debate, and his general political attitudes and opinions. The proclivity of each scientist to dwell at length on a given question or to discuss it summarily was given rather free rein, with minimal guidance by the investigator. The hoped-for benefit of this "client centered" interview was to maximize communication between the interviewer and the respondent. The cost of this procedure was primarily that not every question was answered by each scientist. This accounts for what will appear to be fluctuating numbers of respondents in the tables that follow.

The Cast

The 152 scientists who were active participants in the ABM issue can by no means be considered as representative of the scientific community in general, as a quick comparison to the 1968 National Register of Scientific and Technical Personnel[11] will demonstrate. The Register reports three-fifths of its registrants in the physical and mathematical sciences, one-fifth in the life sciences, and the remainder in the behavioral and social sciences. Our sample consists of 91% (n = 138) scientists in the physical, mathematical, and life sciences and 9% (n = 14) in the social sciences. Of the 297,942 scientists in the Register, 9% were women; there are no women in the ABM sample.

As was the case in other recent occasions when American scientists became politically active, such as the movements to establish civilian control over the Atomic Energy Commission and to establish the National Science Foundation, among the activists one finds a greater proportion of physicists than scientists from any other field.

Industry employed 32% of the scientists in the National Register but only 16% of the scientists in the ABM sample. The proportion of scientists employed in educational institutions is 40% for the National Register and 60% for the ABM group; and those in the federal government is 13% for the Register and 18% for the ABM sample. Those employed by nonprofit organizations is 4% and 6%, respectively. Compared to the scientists in the National Register, the ABM scientists were older (median age age 46.9 vs. 38.0), contained more Nobel Laureates (4% vs. 0.02%), more members of the National Academy (18% vs. 0.3%), and published more articles cited in the Science Citation Index (mean of 60 vs. 6).

Membership on advisory committees on the ABM was one of the criteria for selection for the study. Thus, it should not be surprising that 16% of the ABM sample have been members of the President's Science Advisory Committee (PSAC) or the PSAC Military Strategic Panel. Likewise, 16% have been members of the Defense Science Board (DSB) or Department of Defense (DOD) Task Forces on ABM, with considerable overlap in membership on these advisory committees.

In addition to comparing the ABM scientists to the greater scientific community, we need to place them in the larger perspective of political participation in general. A wide variety of research findings point to a patterning or clustering characteristic of political activity.[12] According to Milbrath, "The variables that correlate with a specific political act tend to correlate with other political acts as well . . . for example, higher socio-economic status (SES) is positively associated with increased likelihood of participation in many different political acts; higher SES persons are more likely to vote, attend meetings, join a party, campaign, and so forth."[13]

A second characteristic of participation, according to Milbrath, is that participation can be conceptualized as being cumulative.[14] Persons who engage in the gladiatorial behaviors are very likely to perform the transitional and spectator activities as well.

In this light, we should expect the ABM scientists—who were in this instance politically active almost by definition—to have also engaged in a variety of other political and nonpolitical activities. The data would seem to support this proposition. Seventy-two percent of the respondents (n = 101) indicated that they had spent equal or greater amounts of time on other issues not directly related to their professional work, in addition to the ABM. Another 4% had spent half as much time on other outside activities as they had on the ABM.

To complete this initial examination of the ABM gladiators, we need to separate the antagonists, to place the combatants in their respective sides of the arena. The possibilities for antiballistic missile defense might properly be viewed as existing on a continuum, starting with (1) no defenses at all; proceeding to (2) an upgraded or modernized air defense system: through (3) a very limited deployment (either to protect a few Minutemen bases or perhaps to protect the national capital and/or command and control headquarters*); and (4) a light area defense designed to provide country-wide protection against a light attack or an accidental launch; to the upper end of the spectrum (5) a heavy urban defense, intended to provide protection of the entire country.

---

* Such defenses, particularly when the points to be defended are relatively invulnerable underground facilities, are commonly referred to as "hardsite" or "hard-point" defenses.

One would then expect to find scientists arraying themselves along this continuum. As Table 1 indicates, however, the middle ground was not heavily populated. The majority of scientists took an extreme position.

Table 1 raises the question of whether knowing the lineup of scientists supporting or opposing deployment of an ABM in general, one could make a prediction about their stand on a light area defense (Sentinel), hard-point defense (Safeguard), or on a heavy urban defense. Table 2 shows the scientists' stand on a general ABM deployment as the independent variable.

Table 1. Scientists' Attitudes toward ABM Systems[*]

	ABM in General (%)	Hard-Point Defense (%)	Light Area Defense (%)	Urban Defense (%)
Oppose Completely	62	53	66	78
Oppose with Qualification	8	12	6	1
Neutral	2	3	1	2
Favor with Qualification	4	8	3	3
Favor Completely	22	21	21	9
Don't Know	—	—	—	1
Not Answered or Not Asked	2	3	3	7

* On this and all following tables, percentages may not add to 100 because of rounding.

Table 2. Scientists' Attitudes toward ABM in General versus Attitudes toward Hard-Point ABM

	(n = 148)				
	Oppose Completely (%)	Oppose with Qualification (%)	Neutral (%)	Favor with Qualification (%)	Favor (%)
Oppose Completely	54	8	—	—	—
Oppose with Qualification	—	3	1	3	—
Neutral	—	—	2	—	—
Favor with Qualification	—	—	—	3	1
Favor Completely	—	1	—	2	20

The measure of association, gamma, has a value of 0.99. In this and all following tables where it appears, the statistic is interpreted as a measure of the
proportional reduction in error of estimation (P-R-E) made possible by the relationship. [15] Thus, by knowing a scientist's stand on ABM in general, our error
in predicting his stand on hard-point defense is reduced by 0.99. Gamma values
of 0.99 and 0.97 were likewise obtained by comparing attitudes toward ABM in
general against attitudes toward light area and heavy urban defenses.

Because these variables are so closely correlated, the single measure, attitude
toward ABM in general, will be the one used throughout the study, unless otherwise
noted, to designate pro- and anti-ABM scientists. Scientists who oppose ABM completely or with some qualification, we shall refer to as anti-ABM scientists (n = 106).
Scientists who support ABM completely or with some qualification, we shall designate as pro-ABM scientists (n = 40). (The missing six are those scientists who
either professed to be completely neutral or would not give their stand.)

The close relationship among attitudes toward all kinds of ABM systems would
seem to lend credence to our analogy of "combatants lining up on one or the other
side." Among the participant scientists, there were few neutrals. The ABM was
a symbolic issue to be fought out between "the Good Guys and the Bad Guys."

THE CALL TO ACTION

What causes an inanimate object, consisting of radars, computers, and missiles,
to become a symbolic issue for scientists, a cause célèbre? Foremost among the
many interrelated components involved in the making of this issue was the phenomenon of the scientists' entrance into the political arena in behalf of, or in opposition
to, the system. There emerges an obvious circularity: the ABM became an issue
because the scientists helped to turn it into one by their public participation. The
scientists latched onto the ABM in part because it was a question with which they
could turn public attention to the greater issues as they perceived them: arms
control for some, Russian parity or supremacy in strategic armaments for others,
strengthening or legitimizing the efforts of the Senate to reassert its role in military
and foreign policy for yet others.

Self-Perceived Motivations

In our analysis of the scientists' motivations, it is important to distinguish between
"establishment" scientists, those who have held high-level government positions or
have been members of prestigious advisory groups such as PSAC, DSB, or their
panels, whom we shall call "Inners" (n = 59), and all others, whom we shall term
"Outers" (n = 93).

One of the opening questions of each interview was "What motivated you to become active and involved in the ABM issue?" As is to be expected, a wide variety of answers was elicited by this open-ended query, with many scientists indicating several reasons. Table 3 shows the number of scientists listing each category of response as their primary or secondary motivation for participation in the ABM issue.

The Inners The single, most frequently given reason among all the scientists was "personal involvement with the ABM issue for many years." One might expect this answer to come predominantly from the Inners, and, as Table 4 shows, this is indeed the case.

Table 3. Frequency of Reasons for Scientists' Participation[*]

Answers having to do with

personal involvement in the ABM, as consultants, doing analyses, working on the system	68
arms race issues or international aspects of an ABM system	49
being asked or recruited by others	39
priorities, misuse of money	20
the quality of the debate. "It was too emotional" or "It was not educational or technical enough."	18
siting of missiles near large cities	15
technical shortcomings of the system	11
the other side, "They were too vociferous."	10
the social responsibility of scientists "to speak out," "to use their knowledge in a useful way."	9
peace and/or the war in Vietnam	6
the public."The public was being misled."	5
the military-industrial complex	4

* Tabulation is based on the first and second reasons mentioned by 147 scientists for their participation.

Table 4. Primary Motivation for Involvement of ABM Scientists:
Inners and Outers

Reason	Inners (%)		Outers (%)
Personal Involvement	79	(n = 147)	17
Arms Race Issues	4		34
Recruited by Others	7		11
Priorities	—		10
Quality of the Debate	2		4
Siting in Cities	—		10
Technical Shortcomings	—		2
The Other Side	7		1
Social Responsibility of Scientists	—		3
Peace, Vietnam War	—		3
The Public Misled	2		2
The Military-Industrial Complex	—		4

The underlying question, however, was what motivated these establishment scientists to shed their customary cloaks of quiet persuasion with the executive departments to done the more visible and strident coats of arousing the public. Several of these scientists expressed a deep feeling of betrayal.

In January 1967, each of the former and present presidential science advisors and directors of Defense Research and Engineering (Drs. Brown, Foster, Hornig, Killian, Kistiakowsky, Wiesner, and York) was invited by Secretary of Defense McNamara to present his views on a proposed ABM deployment to the president. One of these men recounted the story in the following manner:

Late in 1966, I was called to Mr. McNamara's office and was given a memorandum by Secretary McNamara which advocated limited deployment of the Nike-X ABM. I was asked to prepare my views on the subject and present them at a White House meeting in January. I took the memorandum back with me and studied it. The document mostly refuted the use of antiballistic missiles but in the end recommended deployment.

Shortly before the meeting took place in the Cabinet Room of the White House, we met in Dr. Hornig's office and discovered that each one of us had formulated views strongly opposing deployment. We decided quickly that each of us would present personal, uncoordinated views in our five-minute presentations to the President.

The meeting began with the chairman of the Joint Chiefs of Staff, General Earle Wheeler, recommending deployment of the Nike-X system.

Killian, Kistiakowsky, Wiesner, and York then each presented their arguments opposing deployment. Wiesner recalls that he voiced his own opinion that the army would only back a proposed light deployment (Sentinel) if it were a precursor to a larger thick deployment. President Johnson asked, "Is that right, Buzz?" To which General Wheeler replied, "Yes, Sir."

The scientists left the White House thinking they had voiced their independent opposition to deployment of an ABM as clearly and firmly as they were able.

In his September speech, Secretary McNamara, in the view of one of the participants, misinterpreted the conclusions of the science advisors when he stated that

The four prominent scientists— men who have served with distinction as the Science Advisors to Presidents Eisenhower, Kennedy, and Johnson, and the three outstanding men who have served as Directors of Research and Engineering to three Secretaries of Defense—have unanimously recommended against the deployment of an ABM system designed to protect our population against a Soviet attack. (Emphasis added)[16]

Killian recalls that, "Mr. McNamara called each of us in advance of his San Francisco speech to assure us he was not seeking to give an indication that we were supporting his changed position."

Despite these reassurances, however, at least some of these distinguished scientists felt betrayed and were puzzled that Secretary McNamara "stated a different position in his speech than that we had understood him to hold." Said one of them, "We were opposed to all kinds of ABM, not just a heavy anti-Soviet deployment."

Kistiakowsky aired his views on this explicitly when he wired Senator John Cooper, "Secretary McNamara's speech of September 18, 1967, might have given unintentionally the impression that I endorse immediate start of deployment of Sentinel thin ABM system. I do not...."[17]

If a feeling of betrayal was the dominant motivation for some of the anti-ABM Inners, what prompted some of the other leaders of science to become vocal opponents of a system they had been content to oppose quietly hitherto? Several expressed dismay, not about advice misrepresented or unheeded but about advice not sought or asked for at all. When queried about his reaction to the McNamara speech,

Hans Bethe, member of the PSAC Military Strategic Panel, replied he was "pro-
foundly shocked." "It was," he said, "one time PSAC was not asked for advice."

Similarly, with regard to President Nixon's March 1969 Safeguard decision,
W. K. H. Panofsky, another member of the same committee, stated, "The Safeguard ABM system has not been discussed by any of the science advisory bodies
which you mention (i. e. , PSAC or its panels) before the decision was communicated
to the public. "

To some extent, one could say that these Inner scientists took it almost as a
personal affront that, despite the hundreds of hours they had spent detailing the
arguments against the deployment of the Sentinel-Safeguard ABM, it kept reappear-
ing in one guise or another. It should be noted that these particular experts (Bethe,
Garwin, Panofsky) did not oppose antimissile defenses, per se, but felt strongly
that the particular components of Nike-X-Sentinel-Safeguard with the very large,
soft, and expensive missile site radars were unsuited and ill designed for the
purposes proposed.

For other scientists, the change of administration in Washington brought about
their departure from the government and with it, freedom to assume a public advocacy position. "It had every conceivable thing wrong with it, " said one, a chemist.
"If you really wanted to change the way of doing things in the Pentagon this was an
ideal issue. It was lousy technically, and you could get lots of public attention. "

The primary motivations for the anti-ABM Inners can thus be explained partially
by ire over advice unheeded or unsolicited and partially by their exodus from the
government. But, what prompted the pro-ABM establishment scientists, who also
customarily stayed behind the scenes, to enter the public arena?

Table 5 shows the primary reason given by all Inner scientists (pro- and anti-ABM)
for becoming participants.

From Table 5 we conclude that the preponderant majority of the Inner scientists
became ABM participants precisely because they were Inners: they had been deeply
involved with the ABM question for many years. This would indicate that the events
of 1968-1969 did not activate many additional Inners. Forty-eight of the 59 Inners
were already immersed in the ABM by the time the decision to deploy was announced
in 1967.

If we look at the following three closely related answers as a prime reason for
becoming activist: (1) "I felt the public was being misled, " (2) "the quality of the
debate was so low, " and (3) "the other side was too vociferous, " we find 23% of the

Table 5. Primary Motivation for Involvement of Inner Scientists

	Pro-ABM (n = 27) (%)	Anti-ABM (n = 26) (%)
Personal Involvement	74	81
Arms Race Issues	—	8
Recruited by Others	4	12
Priorities	—	—
Quality of the Debate	4	—
Siting in Cities	—	—
Technical Shortcomings	—	—
The Other Side	15	—
Social Responsibility of Scientists	—	—
Peace, Vietnam War	—	—
The Public Misled	4	—
The Military-Industrial Complex	—	—

Inner pro-ABM scientists felt that these motivations were central to their shedding their anonymity, while none of the anti-ABM Inners mentioned any of these reasons as their prime motivations. Typical comments of the pro-ABMers were: "I decided to go public because my colleagues were behaving so abominably"; "I was annoyed at the debate in the sense that it was uneven and unfair"; "I didn't think the vocal opponents were intellectually honest."

In summary, for those Inners who became participants after the deployment decision, the anti-ABM Inners became activists either because they were recruited into the fray by their colleagues or because of their concern with the arms race. Pro-ABM Inners became participants after the deployment decision both in reaction to the ongoing debate as they perceived it and to promote policies they favored. The Outers Recognizing that the numbers we are dealing with are quite small, we can still see from Table 6 that for the Outer scientists, a strong stimulus for the pro-ABM side was also the emphatic feeling that the vocal anti-ABM scientists were "the Bad Guys." Said a California physicist, "I recognized it as a cause blown out

Table 6. Primary Motivation for Involvement of Outer Scientists

	Pro-ABM (n = 13) (%)	Anti-ABM (n = 77) (%)
Personal Involvement	31	13
Arms Race Issues	15	38
Recruited by Others	15	10
Priorities	—	12
Quality of the Debate	15	3
Siting in Cities	—	10
Technical Shortcomings	—	3
The Other Side	8	—
Social Responsibility of Scientists	—	5
Peace, Vietnam War	—	4
The Public Misled	15	—
The Military-Industrial Complex	—	3

of control by leftists." From an Argonne Laboratory scientist, ". . . the conviction that the most vociferous of the physicists in opposition were using their reputation in physics to influence opinions in matters where they had little understanding and less competence."

From Table 6 we see that 31% of the pro-ABM Outers, as compared to only 13% of the anti-ABM Outers, cited personal involvement as their primary motivation. Of the 40 pro-ABM scientists, only 14 became active after the deployment decision; 65% of the pro-ABMers were already involved prior to that time. Of the anti-ABM scientists, however, 74 became activated after September 1967, compared to 35 who were involved before.

According to Table 6, a greater variety of motivations was expressed by Outer scientists opposed to ABM. A partial explanation for the diversity of responses is that the activation of these scientists was not simultaneous but occurred over a period of several years.

Among the earliest nonestablishment science opponents was a small group of grad-uate students and professors at the University of Washington. A young graduate

student, Newell Mack, appears to have been the "catalytic agent" for initial concern with the ABM. According to Mack, he became interested in problems associated with nuclear war in the early 1960s and by 1965 was worrying about the possibility of an anti-Chinese ABM deployment in the United States. In the fall of 1965,
Mack spoke to Senator George McGovern about the possibility of raising the ABM
as an issue in the Senate and also discussed the problem with David Inglis, another
early opponent of the system. [18]

By July 1967, Mack had prepared an outline "Missile Defense: A First Glance,"
which he distributed to scientists attending a talk given at the University of Washington by Hans Bethe. In this prescient four-page paper, Mack raised many of the
questions that would be debated at great length two years later, such as the hardness of Minuteman silos, the number, payload, and reliability of Soviet missiles,
arms race implications of deployment, and eventual costs of a deployed antiballistic
missile system. [19]

While Bethe was at the university, the group discussed with him the possibility
of producing an anthology of ABM writings. Bethe agreed, "such an anthology would
be very timely and would help to form public opinion on this important issue." [20]
However, with the announcement of the decision to deploy in September of that year,
plans for the anthology were abandoned.

On November 15, 1967, the army announced the first ten geographical areas to be
surveyed as possible locations for the area defense system. These were listed as:
Albany, Georgia; Chicago, Illinois; Dallas, Texas; Grand Forks Air Force Base,
North Dakota; New York, New York; Oahu, Hawaii; Salt Lake City, Utah; Seattle,
Washington; Boston, Massachusetts; and Detroit, Michigan. [21]

When he realized that the proposed missile site for Seattle was to be at Fort Lawton, within the city limits, Mack began to wonder where the other missile sites for
the proposed area defense would be located. On December 19, 1967, Mack wrote to
Bethe: "I don't know whether Sprints are to be placed so close to other cities tenta-
tively chosen as possible locations for Sentinel bases....If Sprint missiles are to be
placed in or near these cities, then the 'thin' defense begins to look like a destablizing
'thick' defense." [22]

In real detective fashion, he began to ferret out the information. Mack wrote to the
senators of each state where a site was announced, requesting a precise location of
the radar sites. Senator John Tower replied, "This is information that I do not feel
at liberty to divulge. I would suggest that you communicate with the Army Air Defense
Command." [23] Other senators, apparently not knowing the precise locations themselves,
forwarded Mack's letter to the Department of the Army.

Colonel Raymond T. Reid, of the Office of Legislative Liaison wrote to Senator Herman Talmadge of Georgia on April 22, 1968, that "a precise location for the radar has not been established at this time." He went on to add, "It may be of interest to you that a similar request from Mr. Mack has been made upon individual Senators of most states announced as a potential Sentinel location."[24] Senator Talmadge forwarded this on to Mack.

The Secretary of the ABM Committee of the Seattle Association of Scientists, as they were now called, wrote to newspapers in each locality named as a site, requesting additional information. By the early summer of 1968, Mack pinpointed the precise locations for seven of the proposed sites and came to the conclusion that "the sites being considered for battery locations are so close to major cities as to defend the cities."[25]

At Argonne National Laboratory, too, the chance discovery that the army was surveying proposed sites for Spartan missiles in their vicinity was the trigger for activation. A group of Argonne scientists, members of the local Federation of American Scientists (FAS), had been holding weekly lunch-table discussions about the arms race and arms control. As one physicist explained, "Due to David Inglis' momentum, the FAS group became opposed to increases in nuclear armament." When the local Sentinel sites were discovered in October 1968, the latent concern with arms control sprang to the foreground. One of the scientists expressed it in this way:

It seemed this was an issue where something could be accomplished at a local level and bring the question of the arms race home to people in a much more immediate way than is the case usually. Forcing people to think about these issues in such an immediate local context was useful. I felt that there I might do some good by participating other than just easing my own conscience which tends to be my attitude towards such activities in general.

More Outer anti-ABMers (38%) mentioned arms race implications of ABM as the primary reason for their concern and activity in the ABM issue. The related topic of national priorities and the need to cut military spending was given by 12% of the anti-ABM Outers and was not mentioned by the pro-ABM scientists.

The issue of the social responsibility of scientists was mentioned by 5% of the anti-ABMers. The February 1969 American Physical Society meeting in New York and the concurrent, but independent, formation of Scientists and Engineers for Social and Political Action (SESPA) drew more physicists into the activists' ranks.

Typical responses of the scientists, who mentioned technical shortcomings of the system as their primary reason for involvement, ranged from, "ABM is an example of expensive misapplied exotic technology. It won't work," (a Bell Telephone Laboratory physicist) to, "I was and still am opposed to the ABM on general, almost thermodynamical grounds as another form of pollution of our country," (physicist at the University of New Mexico), and, "It was a conspiracy to foist off upon the public a system which was a glorified military WPA project. I became convinced that the state of the art was still not adequate," (applied mathematician, University of Washington).

If any pattern can be discerned from such a variegated and disparate set of self-perceived motivations, it would seem to be as follows:

1. Some of the most eminent advisors despaired of being able to continue exerting quiet influence from within the Executive Branch. "I decided to try to participate mildly through the Congress instead," explained a former special assistant to the president for science and technology.

2. As the prominent scientists began to voice their opposition, eminent proponents became disturbed and, in the words of a chemistry Nobelist, "tried to furnish evidence that not all scientists were against it."

3. Perhaps encouraged by this example and certainly influenced by the opportunity of having an issue with which they could arouse public attention (missiles in the backyard), scores of scientists, unknown publicly and professionally, entered the political arena to do battle with the ABM.

4. This evoked a predictable reaction from scientists supporting deployment. While characterizing their opponents as "a vocal minority of scientists whose training, judgment and experience did not fit them for the making of this practical decision," they proceeded to reach just such "practical decisions" (to support publicly an antimissile defense) as well.

Why ABM of All Things?

In addition to questioning the scientists on why they personally became involved in the ABM, the study also probed the broader area of why the antiballistic missile, rather than Multiple Independently Targetable Re-entry Vehicles (MIRVs) or the B-1 bomber, became the issue upon which so many of them chose to take a stand.

When the question was phrased this way: "Why do you think scientists picked the ABM?" rather than: "Why did you become involved in the ABM issue?" the most frequently given response was that the ABM was a ready-made political issue. That is, the siting of missiles in metropolitan areas gave the scientists a cause in which widespread coalitions rapidly became possible.

While both pro- and anti-ABM scientists agreed that the public outcry against "missiles in the backyard" was the critical ingredient in snowballing ABM into an issue, the role of scientists as the instigators, or as followers, was perceived differently by the two sides. The pro-ABM scientists saw their scientific colleagues as "stirring up the public," "creating the phony issue of accidents." Their general feeling was that, if only these scientists had not aroused the citizenry, all would have been well. The anti-ABMers saw their role as one of "educating and informing the public," and, moreover, latching on to the ABM after it became an issue, rather than creating the issue.

In contrast to ABM, MIRVs and B-1 bombers would not have consumed any suburban property. Elizabeth Drew, writing about the ABM debate in The Atlantic, concluded that, "in part, the intensity of the anti-ABM sentiment was a fluke, provoked by good old American feelings about real estate."[26]

The answer that ABM has serious technical uncertainties or difficulties was the second most frequent answer of the scientists. Many felt that MIRV was "technically sweet" in J. Robert Oppenheimer's famous phrase. "One can't say that MIRV won't work," explained a former PSAC chemist. "Therefore, it could only be opposed on purely political grounds. But, I have both technical and political reservations about ABM," he added. Even pro-ABM scientists agreed that "there are real technical uncertainties with ABM; that is not the case with MIRV."

Another contributory factor of why ABM and not MIRV or the B-1 bomber became the issue, cited by the respondents, was the qualitative "go" or "no-go" nature of ABM compared to the incremental approach possible with MIRV. As Jerome Wiesner, former presidential science advisor, stated: "MIRV grows along, it is not a 'yes' or 'no' commitment. You can put a MIRV on one missile and have a MIRV deployment." With MIRVs, initial deployment costs are discussed in terms of millions of dollars; initial ABM deployment costs are discussed only in billions of dollars. Even though MIRVs on Poseidon and Minuteman III will eventually cost billions of dollars, "the vastly greater sums of money involved in ABM" were mentioned frequently by the scientists.

One aspect on which there was wide agreement by the opposing sides was that there was little discussion of, or opposition to, MIRV deployment within and outside of the government, compared to the ABM. Both pro- and anti-ABM participants agreed that

1. The ABM had been debated within government for ten years.
2. Arguments in public tend to reflect arguments in private, within the administration.

Still another explanation offered as to why the scientists did not oppose the B-1 bomber was that, "Physicists like something new. One bomber is just another bomber, but ABM was something entirely different. It was a new ball game."

In summary, then, the scientists had a variety of explanations as to why ABM became such an important issue. Among these were the political opportunities that arose from the placement of missile sites in urban areas, the technical uncertainties and the alleged far greater costs of ABM compared to MIRV, the greater interest aroused by a qualitatively different weapons system and by the existence of long-standing disagreements within the government.

SCIENTISTS DO "THEIR THING"

Stephen Dedijer, of Lund University in Sweden, has suggested a Brinell classification scheme for the physical and social sciences according to the firmness or hardness of their data.[27] The motivations we discussed in the previous section are subjective, open to differing interpretations, and would be, according to Dedijer, "whip cream soft." The actions of the ABM scientists are, however, firmer data and move us up a little on Dedijer's continuum. In this section, we shall examine what activities the scientists engaged in, in their efforts to effectuate or prevent deployment of an antiballistic missile system.

Advisory Panels

The scientists' earliest involvement with the ABM consisted of serving as members of advisory panels. As early as May 1946, an army study group, the Stillwell Board, recommended the development of an antimissile defense.[28] A still classified Lincoln Laboratory Summer Study of 1952 which, according to some scientists on it, was primarily looking at antiaircraft defenses, also concluded that antiballistic missile defenses were possible. After a hiatus of several years, the ABM reemerged as Nike-Zeus, in a feasibility study conducted for the army by the Bell Telephone Laboratories and completed in September 1956.[29]

Almost as soon as the army recommended deployment of the Zeus system in 1958, the ABM became a controversial issue and scientists began to line up on one side or the other.[30] With President Eisenhower moving his Science Advisory Committee (PSAC) into the White House and creating the office of Special Assistant for Science and Technology in 1957, the establishment of the Advanced Research Projects Agency (ARPA) in 1958, and the Directorate of Defense Research and Engineering (DDR&E) also in 1958, many of the scientists who were to become major players in the 1969-1970 ABM drama began their involvement with the ABM issue.

Fifty-seven (36%) of the ABM scientists eventually served on advisory panels or task forces concerned with missile defenses. The most important of these have been PSAC and its Military Strategic Panel, and the Defense Science Board (DSB) and task forces established by the Defense Department (DOD) on which 42 scientists served. (Nine scientists served on both PSAC and DOD panels.) While there usually was at least one proponent of deployment on the PSAC panels and occasionally opponents of deployment on a DOD panel, in general, the members of PSAC and its panels opposed deployment while the members of DSB or its panels were in favor of deployment. This is shown in Table 7.

There are at least three plausible hypotheses to account for such a polarization. One would be that the scientists' judgments on the feasibility or desirability of an ABM were arrived at independently and were known to the policy maker, who then selected them for their views as well as for their expertise.

It is useful to consider in this respect both the PSAC and DOD panels. Many of the scientists who were on the PSAC Military Strategic panels were, in the late 1950s and early 1960s, advocates of stressing defensive over offensive strategies and were, therefore, favorably inclined toward development of an ABM. By about 1962-1963, some of them changed their views, due to technological advances in offensive missiles and to their worries about the arms race implications of a deployed system. Since they were reappointed to PSAC despite their changing views, the hypothesis would seem to be disconfirmed.

However, it should be pointed out that the reappointments were not necessarily made by the same man who made the original nomination. The chairmen of PSAC panels are named by the chairman of PSAC, who by tradition has been the president's science advisor. From 1957 to the present, there have been six chairmen of PSAC.

Table 7. Stand on ABM of Members of PSAC and PSAC/ABM Panels and Members of DSB and DOD Advisory Panels at Time of ABM Debate

	PSAC and/or PSAC Panels (n = 25) (%)	DSB and/or DOD Panels (n = 26) (%)
Anti-ABM	76	23
Pro-ABM	24	77

It is also possible that the incumbent PSAC chairman changed his views on anti-missile defenses in step with those of the panel chairman he reappointed.

John Foster, director of Defense Research and Engineering, who appointed many of the DOD panel members said quite frankly that "a certain amount of partiality is built into any advisory body in the selection of its members." But, Foster felt that, if one only chose panel members who were known to agree with the agency's views, they would not be performing their primary function of examining the agency's proposals with a critical and evaluative eye.

A related, but somewhat separate, hypothesis would be that the policy maker chooses panel members who have other connections with the agency and who may therefore not really be free to dissent from the department's primary objectives.

Questions dealing with the source of funding of the scientists' research were included in the interviews. For the entire sample of ABM scientists, 35% reported that some or all of their research was funded by the Defense Department. (However, it should be noted that information regarding funding sources was obtained from only 98 respondents, 65%.)

Table 8 looks at the subpopulation of those ABM scientists who were members of PSAC or DOD advisory panels.*

With 86% of DOD panel members funded by the Defense Department and 66% of DOD panel members working for the government or industry at the present time, compared to 47% and 20%, respectively, for PSAC panel members, the data would seem to confirm the hypothesis that DOD panel members tend to be scientists who have funding or employment ties to the Defense Department.

It is interesting to note that deputy secretary of Defense David Packard has stated that, "I do not consider that when you are involved with scientific matters it is important whether you have people outside the Defense Department or not. Scientists, to me, are objective about such matters."[31]

Finally, one could also hypothesize that the homogeneity of the respective views of PSAC and DOD panel members is due to the effect such a group exerts on its members. Solomon Asch, in his experiments on the effect of group pressure, found that a substantial minority (37%) of his subjects would yield to majority opinion, modifying their judgments in accordance with the majority even when the group's judgments were clearly perceived to be contrary to fact.[32]

---

*It should be noted that six members of these panels are university administrators who do not engage in research themselves. Although in each case it is quite likely that their institutions receive Defense Department funds for research, these six men were not included in the last item of Table 8.

Table 8. Employment and Defense Department Funding for DOD and PSAC
Panel Members

	DSB/DOD Panel (n = 26) (%)	PSAC Panel (n = 25) (%)
Employed by Government	20	8
Employed by Industry	48	12
Employed by Academic	19	76
Employed by Nonprofit	15	4
Funded by Defense Department	86 (n = 21)	47 (n = 15)

Accordingly, one could expect pro-ABM scientists on PSAC gradually to shift to
a more negative viewpoint, submitting indirectly perhaps to the group norm and
anti-ABM scientists on DOD panels to adopt, more and more, a position favoring
deployment. In keeping with this hypothesis, the clearly dominant anti-ABM feelings of the majority of the PSAC panelists did seem to serve to inhibit the expression if not the maintenance of an opposing viewpoint of at least one panel member
interviewed. On the other hand, Richard Latter, of the Rand Corporation, a member
of the PSAC Military Strategic Panel, seems to have always withstood whatever
pressures that group may have exerted and has expressed his pro-ABM views
clearly and vigorously, according to other panel members.

In summarizing the three hypotheses, the data seem to the author to be suggestive
rather than conclusive evidence for confirming the latter two hypotheses.

Since none of the hypotheses is flattering to the scientists' own view of their independence, it would seem that the anomalous members deserve a closer look. By
anomalous members, we refer to those scientists on DOD panels who were opposed
to ABM deployment or those who were in favor of ABM and were members of PSAC
panels. If scientists served on both PSAC and DOD panels, then they must have been
anomalous members on one panel or the other. With three exceptions, all of the
anomalous panel members are those and only those scientists whose memberships
overlap, that is, they were members of both PSAC and DOD panels.

If we now ask how these overlapping or anomalous scientists are selected for
membership on the advisory committees, the following hypotheses can be considered:
HYPOTHESIS 1:
The scientists are so prominent that they cannot be excluded, that is, the credibility
or legitimacy of a panel which did not include them would be questioned. Looking at
our three indicators of eminence—being a Nobelist, membership in the National
Academy of Sciences, and scientific citation rate—we find no statistically significant
differences between scientists who did and did not overlap.

HYPOTHESIS 2:

The scientists are first appointed to the committee where they are in the minority or anomalous position. Then later they become a member of a panel where they are in agreement with the majority views. This also does not seem to be borne out by the data.

HYPOTHESIS 3:

The stands of the scientists are known to the decision maker and the anomalous appointment is made to "balance" or "legitimize" the advisory group. This seems well supported by the available data. This hypothesis is further strengthened when we refer back to the strong polarization shown in Table 7. In going over panels that have examined the ABM issue since 1967, in only one instance could one speak of such a phenomenon as a "nearly evenly divided" panel.

The mantle of executive privilege makes it extremely difficult to obtain reliable information about the performance, scope, and recommendations of these advisory groups. The following information, which has been obtained by a variety of techniques including interviews, deduction, and cajolery, gives a glimpse into the proceedings within these inner sanctums sanctorum.[33]

To begin, it may be useful to contrast the Defense Science Board and Defense Department task forces with PSAC and PSAC panels. DOD panelists appeared to take a narrower view of their responsibilities and tended to view their position as less independent. As one DOD panel member said, "We really work for DDR&E and John Foster. We respond to his concerns." This same member stressed that Foster utilized the Defense Science Board much more than his predecessor Harold Brown did. However, the process may be circular; a panel member may suggest something to DDR&E which then requests a study.

PSAC panels are more likely to initiate avenues of inquiry and originate studies for themselves. DOD panels have a specific charge spelled out to them. PSAC panels are almost always chaired by a PSAC member and the panels report back to PSAC and to the president's science advisor. DOD Task Forces are often chaired by a nonmember of the Defense Science Board and report back to the appointing agency, mostly DDR&E, not to the Defense Science Board. DOD panels are more ad hoc, while PSAC panels, particularly in the case of the Military Strategic Panel, continued for many years. In both DOD and PSAC panels, there is considerable continuity of membership and often overlap between committees as well. For example, Richard Latter in the years 1967-1971 was a member of the PSAC Military Strategic Panel, was chairman of a 1967 DOD Task Force on Ballistic Missile Defense, was a member of the Defense Science Board, was chairman of a 1970 DOD Task Force on Missile Defense, and chaired the 1971 DOD Task Force on Hardsite Defense.

The longevity of membership tends to generate a chummy, club like atmosphere. Thus, we learn that the October 1968 meeting of the PSAC Military Strategic Panel was converted into a working session of Scientists and Engineers for Humphrey-Muskie, to the chagrin of at least one panel member who was a Nixon supporter. He viewed with distaste the sight of his colleagues "arranging calling and canvas activities at the expense of the legitimate responsibilities of PSAC." Other panel members, when questioned about this, either stated they had not attended the meeting or could not remember such an activity occurring. While there has been no outright denial of these facts, Richard Garwin wrote in November 25, 1972 Saturday Review that he did not "arrange calling and canvas activities."

The chairman of a 1967 DOD Panel to assess the Chinese ICBM threat stated that panel members worked eighteen hours per day for six weeks in Washington preparing that report. In contrast, the DOD Ad Hoc Group on Safeguard for FY 1971 (the O'Neill Panel) met for a total of three days before submitting their report. * Since the report was publicly available, panel members felt freer to discuss it and this gave us an opportunity to examine it more closely.

The members of the group were Lewis Branscomb, director, National Bureau of Standards; Sidney Drell, professor of physics and deputy director, Stanford Linear Accelerator Center; Marvin Goldberger, professor of physics, Princeton University; William McMillan, professor of chemistry, University of California at Los Angeles; W. S. Melahn, president, System Development Corporation; Lawrence O'Neill, president, Riverside Research Institute (Chairman); Allen Peterson, professor of electrical engineering, Stanford University.

John Foster stated that he "personally picked the list and called them." Some of the panel members were apparently notified by mail with the charge to the committee spelled out. However, two panelists asserted they were notified by telephone on Friday, with the committee scheduled to meet the following Tuesday. Both these scientists were told over the phone that they were being asked to evaluate DOD proposals for Safeguard Phase II and were told that any objections they raised regarding the system would be passed on to the administration along with the recommendations. However, when the panel convened on Tuesday, the charge laid out by Gardiner Tucker, assistant secretary of defense for systems analysis, was to "offer advice on scientific and technical grounds regarding the deployment of additional elements of Safeguard to be funded FY 1971."[34]

---

* Almost the entire O'Neill Report was declassified and inserted in the Congressional Record, August 6, 1970.

One of the two "telephoned" scientists said that he "felt betrayed" when the meeting got under way and the other man stated that "he appreciates now that, if one joins a panel, one must have in writing what the charge to the panel is." Nevertheless, both scientists did "review and approve" the report when it was issued.

Richard Latter, though not a panel member, attended all panel meetings and deliberations. One panelist said he had insisted on Latter being on the panel because "it would increase our credibility with Johnny Foster," but according to another panel member, "Latter was a noisy observer whose views were not reflected in the report." Said this scientist, "I presume Latter was there to give Foster an account of who said what; he was a functionary of J. Foster's."

The first day was devoted to "briefings by the Safeguard Systems Command, the Advanced Ballistic Missile Defense Agency (ABMDA), and representatives of DDR&E."[35] The second day was given to discussion, the definition of points of agreement, and identification of differences in technical judgments.[36] The preparation of working notes for the group's report and the presentation of its advice to officials in DOD occupied the final day.[37]

In its conclusion ("Pablum-like" according to one panel member), the report offered the following comments: "Safeguard Full Phase II is a system embodying compromises intended to enable the system to achieve, to some extent, the three objectives stated by the President. It provides thin area coverage and some defense of Minuteman. It retains the possibility of including Safeguard technology in a dedicated system for Minuteman defense."[38]

In addition, the report included observations on budgetary matters, the relationship between various deployments, and the realization of the several objectives of Safeguard (that is, defense of Minuteman, light defense against Communist China, and coverage against sea-launched ballistic missile attacks).

There the report might have remained, buried under stacks of similar advisory panel reports, forgotten or unheeded. There are times, however, when officials seemingly want to buttress their projects with an aura of legitimacy and expertise. Such appeared to be the case when, as he faced hostile questioning by members of the Subcommittee on Arms Control, International Law and Organization of the Senate Committee on Foreign Relations, on June 4, 1970, John Foster resurrected the O'Neill Panel report.

In answer to Senator Fulbright's question whether "the Defense Department is capable of an objective view of these matters,"[39] Foster pointed out "that he had asked a group of scientists to come together as an ad hoc committee . . . and to review the program." He continued that "the report sent to the Secretary of Defense said this equipment will do the job that the Department of Defense wants to do."[40]

That must have caused the telephone lines to sizzle, for the very next day Sidney Drell, one of the panel members, wrote to Senator Gore, chairman of the sub-committee, "It seems to me that Dr. Foster's remarks indicate that we made re-commendations which, in fact, we did not make."[41]

Another panelist, Marvin Goldberger, indicated, "The report took no such posi-tion."[42] On June 29, both scientists appeared before the subcommittee and vigor-ously denied that the O'Neill Panel had reached the conclusions attributed to it by Foster.

From the timing of the meeting of the O'Neill Panel, it would appear that the panel contributed at most a very marginal or minor input into the decision-making machinery. The panel met on January 6-8, 1970. President Nixon announced the FY 1971 plans for Safeguard at a press conference on January 30, 1970.[43] It would seem that the decision makers who convened the panel had substantially decided what course to recommend, and it appears to this observer doubtful that there would have been any substantial difference in the president's recommendation had the panel never met. One can speculate that, perhaps, if the panel report had pointed out a gross technical blunder or had used less ambivalent and watered-down terminology, the result might have been different, but this must remain speculation.

In sum, scientists have been active as advisors in the ABM issue since the idea of "hitting a missile with a missile" was just that—a concept, a dream, or a night-mare. The advisory panels established within the executive department, both by the President's Science Advisory Committee and by the Department of Defense, have been highly polarized groups with the PSAC panels consistently opposing deployment and the DOD panels favoring deployment. The composition of the panels has been fairly homogeneous, with PSAC panel members coming primarily from the academic world and DOD panel members being mostly industrial and government scientists. The evidence would seem to indicate that the DOD panelists are not entirely free from dependence upon or allegiance to the Defense establishment. That is not to argue, however, that there are no pressures working on PSAC panelists. There assuredly are but, being perhaps of a more subtle nature, are more difficult to recognize.

Friends and Neighbors

Among the other activities the scientists engaged in, speaking, lecturing, and debat-ing about the ABM in their own communities were the actions in which the largest number of them participated. In Seattle, Berkeley, Honolulu, Minneapolis, Detroit, New York, and Boston, more than half (51%) of the ABM scientists "took to the husting" to extol the virutes or vices of ABM systems to women's clubs, Chambers of Commerce, students, bird watchers, or clergymen—to almost any warm body willing to listen, it appeared at times.

While the scientists who served on advisory committees were by definition Inners, the scientists who spoke out in their own communities were predominantly Outers. Of the 76 scientists who lectured or debated the ABM issues before their neighbors and friends, 78% were Outers and only 22% were Inners. What caused these Outer scientists to become activists was, in many instances, the serendipitous nature of the discovery that rockets 55 feet tall, carrying megaton-sized thermonuclear warheads were to be emplanted literally in their backyard. Many elements of the activation of Outer scientists into the ABM issue are illustrated by the Argonne National Laboratory story.

In the flat, square, neat middle-class communities of Chicago's western suburbs, a biweekly community newspaper, Suburban Life, on October 26, 1968 contained a small story about "The Chicago base of the Sentinel Missle sic Air Defense system which will be located either on a portion of the Healy farm land... or west of Westchester."[44]

Since both locations were within eight miles of his own home and Argonne National Laboratory where he was employed as a physicist, John Erskine stopped off at the farm to inquire "what was going on." The drilling rig operator laconically mentioned "something about missiles, you know the big ones." Erskine called the Army Corps of Engineers in Chicago who verified that the diggings were indeed for the Sentinel ABM system.

As a member of the FAS group at Argonne, Erskine had read and discussed with David Inglis (then also at Argonne) his articles[45] opposing the United States' deployment of one. He knew of former Science Advisor Jerome Wiesner's opposition to ABM,[46] and he had read the Richard Garwin-Hans Bethe article detailing technical shortcomings of an ABM.[47] But until that moment, the ABM was simply one of many issues—urban blight, pollution, prejudice, Vietnam—about which Erskine and thousands of other scientists had vague feelings of uneasiness.

Suddenly and dramatically ABM exploded on the collective consciousness of the Inglis-FAS group at Argonne as an issue of immediacy and paramount importance. The scientists were propelled into action.

They scrutinized the record of the previous year's hearings of the Subcommittee on Military Applications of the Joint Committee on Atomic Energy and abstracted them for handy reference. They prepared charts and graphs from The Effects of Nuclear Weapons, an AEC and DOD publication, and they telephoned numerous queries directly to the Huntsville, Alabama Sentinel System Headquarters.

Before three weeks elapsed, two of the Argonne scientists contacted friends in the television and newspaper media, and on November 15 Chicagoans awoke to find two-inch headlines querying "A-Missile Sites in Western Suburbs?"[48] Television, radio and newspaper reporters interviewed the Argonne scientists, and three days later, the scientists now calling themselves "The West Suburban Concerned Scientists" had prepared a position paper discussing the issues of cost, effectiveness, and arms race escalation of ABM. Local government officials in areas adjacent to proposed missile installations were contacted with an offer to send speakers and information packets.[49]

Almost immediately, the public meetings began. The army organized a briefing session for local congressmen, governmental officials, and the public, which the Argonne scientists attended. The questions of safety raised by the scientists prompted one citizen at the meeting to ask how far from Chicago the missile site could be placed without reducing its effectiveness.[50] Colonel William Wray, chief of site operations for the Sentinel System command, said he could not answer the question for security reasons.[51] After the army briefing, the scientists in attendance were again interviewed by the news media and pointed out that many answers to the questions raised were obtainable in the open literature.

Area residents begain circulating petitions to block the development of the defense facility at either of the two proposed sites. Over 350 citizens of Westchester attended what was to be first of many village board hearings that winter. The panel discussion featured Frederick Wedinger, village president, Congressman Harold Collier, and John Erskine and John Schiffer of Argonne.[52] In contrast to the scientists' eagerness to speak at all possible meetings, the army continued its policy of stating that: "It would not be possible for an Army representative to appear at a meeting to discuss the proposed Winchester site inasmuch as much of the material is classified."[53] At the end of the meeting, after a show of hands indicated that, with one or two exceptions, all 350 people opposed the proposed sites, the village board passed a resolution opposing the missile sites.[54]

Three days after this initial victory, Erskine, accompanied by another Argonne

physicist, Roy Ringo, addressed 75 members of the York Woods Community
Association. Now the speaking engagements spread out to La Grange Park, the
Lombard Democratic Club, and a public meeting sponsored by the Glenview League
of Women Voters. All this flurry of activity centered in the western suburbs. Then
suddenly, on December 12, Libertyville, a northern suburb, was selected as the
site for the installation and the action shifted to the northern suburbs.

By now, with one village and county board after another having passed resolutions
objecting to missile sites within their jurisdiction, the army changed its policy and
dispatched a team "ordinarily consisting of two full colonels, a lieutenant-colonel
working the slide projector and a civilian public relations man with a pipe, a Sentinel
tie clasp and an elaborate tape recorder"[55] to counter the Argonne scientists'
offensive.

As the opposition continued to mount, the army strategists decided to shift their
top team and sent John Foster and Lieutenant General Alfred Starbird to an open
briefing in Waukegan, attended, of course, by the indefatigable Argonne scientists.
Foster said that the Sentinel system was designed to be used against a relatively
primitive Chinese attack. "But," Foster continued, "the matter of thick and thin
is one of degree." and he did not know whether or not the present system might be
expanded. [56] This remark did not allay or mitigate the scientists' fears, and their
speaking engagements continued unabated throughout January 1969, with "the soldiers
and the scientists travelling around Lake County like old prize-fighters staging
exhibitions"[57] until on February 6 the new Secretary of Defense Melvin Laird ordered
a halt to further ABM site investigations pending a project review.

For the Argonne scientists, Laird's announcement resulted in a diminution but
not the cessation of their public speaking on the ABM. At the University of Minnesota, in contrast, the announcement of the Safeguard decision in March marked the
beginning of the anti-ABM effort. Here the focus was on small neighborhood gather-
ings of "affluent and conservative suburbanites," according to a professor of chemical engineering. He estimated he spoke to one such "coffee klatsch" per week
throughout the summer of 1969.

Pro-ABM scientists too were speaking out in their hometowns, pointing out that
many cities already had nuclear weapons stored nearby and that there had been no
accidents caused by these. Of the 76 scientists who lectured or debated in their
communities, 12 were pro-ABM and 64 were anti-ABM. Although a few scientists
charged fees for these speaking engagements, mostly the scientists did not seek any
payment for their lecturing forays.

Peregrinations

At least 44 scientists traveled away from their hometowns, sometimes crisscrossing the country to spread the gospel according to Saint Donald or Saint Jerome. Of these, 59% were Outers and 41% were Inners. These two groups, the Inners and the Outers, reported widely differing receptions, audiences, and experiences.

For example, Nobelist George Wald addressed over 2,300 Chicagoans who paid up to $7.50 each to attend a rally sponsored by Chicagoans against the ABM.[58] In the same month an Outer physicist traveled to North Dakota to speak before "about 50 people, only a few of whom were youngsters, at a public forum...at the Municipal Auditorium. The event...raised $53 in donations. Most of that will be applied toward_____'s plane fare. He received no "speaking fee."[59]

Many of the scientists' speaking arrangements were coordinated by the dozens of ad hoc citizens groups formed by both sides. A feast of acronyms seemed to spring up across the country from SCRAM (Sentinel Cities Reject Anti-Missiles) in Seattle; WOMAN (Women Opposed to Missiles and Nuclear Warheads) in Detroit to NO ABMS (Northfielders Opposed to ABM Systems) in Minnesota.

One such organization was the Committee to Maintain a Prudent Defense Policy, cofounded in May 1969 by former Secretary of State Dean Acheson, former Deputy Secretary of Defense Paul Nitze, and Professor Albert Wohlstetter of the University of Chicago. This committee was formed, according to Acheson's invitational letter, "because of their strong concern that the argument had been largely one-sided." They wanted to "contribute to a balanced debate."[60]

An office was established on Connecticut Avenue in Washington, D.C., and was staffed by a secretary and two young men, one a graduate student of Wohlstetter's and the other, a family friend of the Wohlstetters. The staff arranged for appearances by committee members on the "Today" and the David Frost television programs and set up debates between ABM supporters and opponents.

But the primary function of this committee was issuing nine reports or position papers that were distributed to all senators before the vote was taken in August 1969. The first of these, by Wohlstetter, argued that limiting Safeguard to research and development and testing in the Pacific, as was implicit in the Cooper-Hart amendment, would "bear a high cost of symbolism" and be "an assured waste."

Other papers produced by the committee were devoted to countering specific arguments raised by opponents of deployment concerning the vulnerability of the radars, arms race implications, SS 9 deployment, surprise attack, and launch-on-warning.

Whether the particular group's primary purpose was arranging the speaking engagements of the scientists or providing a forum for their ABM papers, the newly forged alliances between peace, farm, labor, business, religious, and political groups and a sizable segment of the scientific community seemed to exude an aura of fellowship and well-being seldom expressed before between "town and gown."

Congressional Appearances

Among the scientists' many ABM activities, perhaps the most important and far-reaching innovation was the appearance of nonadministration scientists testifying before Congress on a weapons system.

The scientists' first contemporary foray in the halls of Congress was the epic struggle for civilian control of the Atomic Energy Commission in 1945-1946. This battle had many similar characteristics to the ABM fight. Among these was the initial opposition within the Congress to hearing scientific testimony in opposition to the administration's proposal. After holding five hours of hearings, Chairman Andrew J. May of the House Military Affairs Committee said that "his committee had heard everyone who asked to testify and closed its hearings because there were no more witnesses."[61] The four witnesses heard were Secretary of War Robert P. Patterson; General Leslie Groves, commanding officer of the Manhattan Project; Vannevar Bush, director of the Office of Scientific Research and Development; and James B. Conant, chairman of the National Defense Research Council.[62]

According to Alice Smith, "Within ten days of its introduction, a significant portion of the American scientific community was in full cry against the May-Johnson Bill."[63] Within a week the hearings were reopened, and a procession of scientists marched to the Capitol to oppose the May-Johnson Bill. But it must be stressed that the May-Johnson Bill dealt with the establishment of the Atomic Energy Commission. It did not deal with a weapons system.

When the storm erupted about developing the hydrogen bomb in 1949-1950, it occurred behind tightly closed doors. Most scientists and certainly the public only became aware of the existence of that raging battle when the transcript of the Oppenheimer hearings was published in 1954. Nongovernment scientists had testified both for and against the Limited Test Ban Treaty in 1963, but the Limited Test Ban was a treaty, a recognized political instrument, not a weapons procurement, a recognized military prerogative.

To understand the extent of the break with time-honored tradition, it is necessary to take a backward glance. Since World War II, the Congress had not challenged the Defense Department seriously on any appropriations.

As recently as 1966, the Military Appropriations Bill sailed through the United
States Senate without so much as a roll-call vote. [64] It was not only in the halls of
Congress that one witnessed virtual unanimity on every national security proposal.
Throughout the land, only a few voices called out for public debate on such issues
as the proposed deployment of an antiballistic missile system.

We know now, with hindsight, that throughout 1967 this proposal was being fiercely
debated within the executive branch, [65] but during the spring and summer of
1967 there was a notable absence of public discussion of the issue. Ralph Lapp, in
addressing the Midwestern Conference of Political Scientists at Purdue University
in April 1967, cried out in solitary protest:

The President has a science advisor, but he has given no public counsel on Nike-X.
The White House has a President's Science Advisory Committee, but it issues no
public statement—no White Paper on Ballistic Missile Defense. The National Academy of Sciences has a Defense Science Board, but it gives no public counsel on
Nike-X. But the science hawks sweep down and urge more arms.[66]

Senator Jacob Javits commented on what he considered to be "the inadequacy of
the national debate which preceded the ABM decision" in a speech he delivered at
New York University on November 6, 1967, seven weeks <u>after</u> McNamara's announcement of deployment. [67]

"It is essential," Javits stated, "that public men, both in and out of government
join in the continuing debate over the need and justification for ABM defense. Now
is the time we need the views and judgments of the nation's best minds. Later,
when we might be irrevocably tied to the ABM roller coaster, their post mortem
dissent will be of little value."[68]

But when the congressional committees held authorization and appropriation hearings on the Sentinel system in 1968, the traditional modus vivendi continued unaltered,
as was brought out in the following exchange between Senators Fulbright and Russell:

Fulbright:
"Did I understand the Senator to say that no witnesses were brought into the hearings
on this matter except administration witnesses?"
Russell:
"We heard all the witnesses who wanted to be heard."[69]

Hans Bethe, George Kistiakowsky, Jerome Wiesner, and Herbert York had tele-
graphed Senator Cooper on July 23: "We believe that it would be wise to delay this
deployment for a year or more."[70] When this telegram and a similar one sent to
Senator Hart signed by Bethe, Carl Kaysen, Kistiakowsky, Arthur Larson, Burke
Marshall, Roswell Gilpatric, Wiesner, and others were referred to Senator Russell,
he replied, "We have been conducting research on this missile for 12 years. These

scientists, every time a bill comes before the Senate, send a telegram saying this will not add to our defense. But at no time has any of them ever asked to appear before the committee. Not a single Senator has ever asked for one of them to appear. But year after year they send in this telegram when the bill is before the Senate."[71]

The entire proceedings of the rare executive session in which the foregoing colloquy occurred, censored by the Department of Defense at the Senate's request, were quietly slipped into the Congressional Record on November 1, 1968, by Senate Democratic leader Mike Mansfield.[72]

Why did not the telegram senders ask to testify publicly in the spring and summer of 1968? Why did not Senators Cooper, Hart, Javits, or Fulbright seek to have them appear as witnesses? In part, the explanation lies in the strong Senate respect for tradition. Donald R. Matthews has pointed out that the United States Senate, as any group or organization whose members interact frequently, has its "unwritten rules of the game, its norms of conduct, its approved manner of behavior. Some things are just not done...."[73] Matthews continues: "The Senators believe, either rightly or wrongly, that without the respect and confidence of their colleagues they can have little influence in the Senate.... The safest way to obtain this respect is to conform to the folkways."[74]

Because this reverence for adherence to the traditional path is so firmly embedded in the folkways of the Senate, knowledgeable outsiders, such as lobbyists, incorporate or internalize the Senate's mores into their own plans and calculations for action. The Senate's respect for tradition is accepted as a given by lobbying organizations.

The Council for a Livable World (CLW), a scientists' lobbying and fund-raising organization, and the Federation of American Scientists (FAS) both were firmly opposed to ABM deployment in 1968. Each organization had issued position papers and arranged for briefings of selected senators and their staffs by prominent scientists opposed to deployment before the Senate voted in the summer of 1968. But according to Tom Halsted, former CLW national director, "We never thought we'd get the chance to have outside scientists appear before Russell's committees." The folkways of the Senate in effect imposed conceptual blinders on the ABM scientists' efforts in those early days.

Second, in the spring and summer of 1968, the voices protesting missile sites remained isolated and largely unheard and unheeded in the country as a whole. There were, it is true, pockets of scientists' and other citizens' anxiety about ABM expressed, but the voices in Seattle, Washington, and Lynnfield, Massachusetts, remained localized and did not attract nationwide attention. The necessary spark for a conflagration was just not present.

Third, 1968 was such a tumultuous year that ABM as an issue simply could not penetrate the public's consciousness. The New Hampshire primary, President Johnson's decision not to seek reelection, Martin Luther King's and Robert Kennedy's assassinations, the debacle at the Chicago convention and the campaign in the fall— these were the events that preempted the public's attention.

Fourth, and finally, we must remember the political affiliations of the scientists. Most of the ABM activists belong to the Democratic Party, and indeed many of them had held government positions under Democratic administrations. Undoubtedly, it was easier for these men to go public after the election.

Late in 1968 and early in 1969, the confluence of many factors brought about the sudden and dramatic birth of ABM as an issue. Suddenly, in the hiatus between the election and the inauguration, the once isolated and muted voices of protest against ABM began to reinforce each other and grow louder. Whatever "critical size" of public clamor is necessary to transform a concern into an issue was reached as the mass media picked up the stories of citizens' protests across the land. That, in turn, provided the necessary impetus for a break with congressional tradition.

Two factors meshed to bring this about. One was that Senator Russell's earlier denunciation of the scientists for their failure to testify was taken as sufficient reason by some for the scientists to be heard in the future. On December 12, 1968, Cameron B. Satterwaite, chairman of the FAS, wrote to Senator Russell: "I have just read in Science . . . that you stated that no scientists outside the government have requested to testify . . . . If this is true, let me correct it at once and request that the FAS . . . be given the opportunity to provide witnesses to testify at any further hearings under the auspices of your committee which deal with the ABM."[75]

Another contributory factor was the relinquishing of the chairmanship of the Senate Armed Services Committee by Senator Russell to Senator John Stennis. Soon after he assumed the chairmanship, Stennis agreed to a request from Senator John Cooper to allow outside witnesses to testify before his committee.[76]

The combination of rising public opposition to siting ABM missiles near large cities, often led by the scientists, the end of Senator Russell's 16-year reign over the Senate Armed Services Committee, a growing concern with military spending and national priorities, and the scientists' reactions to Russell's rebuke all united in the beginning of 1969 to make the break with tradition possible.

The eminence of the men chosen to testify was stressed from the outset. It was heralded that the witnesses "would include all former Science Advisors to U.S. Presidents, Nobel Prize winners in physics and men from some of the major scientific institutions in the nation."[77] "Credentials count," is how one of the scientists, Herbert York, phrased it. "It puts moxie into our remarks!"

The first outside scientists to testify on the ABM appeared on March 6, 1969, before the Subcommittee on International Organization and Disarmament Affairs of the Senate Committee on Foreign Relations (the Gore subcommittee). A steady procession of scientists—opponents and supporters of antiballistic missile deployment—marched before the Gore subcommittee and the House Foreign Affairs Committee in March, the Senate and House Armed Services Committees in April, and the Gore subcommittee again in May and July 1969. Between March 1969 and May 1971, 33 physical and social scientists of our ABM population testified before congressional committees on the ABM.

In most instances the witnesses were selected well in advance of their appearance, and many scientists spent considerable amounts of time preparing their testimony, replete with charts and graphs. But, probably, the most publicized and, according to the scientists themselves, the most influential, was the first appearance of W. K. H. Panofsky before the Gore subcommittee. As with the chance discovery of the proposed missile sites near Argonne, Panofsky's initial appearance seemed due to Lady Luck or Lady Fate, depending on one's point of view.

According to Panofsky, he was in Washington testifying before the Joint Committee on Atomic Engergy on the subject of collaboration with the Russians in high-energy physics. Before the session at which he was to testify, Panofsky lunched with a staff member who had been present at a meeting of the Gore subcommittee that very morning. He told Panofsky that his name had come up during the hearing. Deputy Secretary of Defense David Packard had been asked by Senator Fulbright; "Did you go outside and employ any independent people to analyze the feasibility and advisability, the wisdom of this program?"[78] Packard had replied: "The review utilized the full staff of the Defense Department and those people that the Department had utilized for scientific evaluation. In addition to that, I have talked to some scientific people on my own about the matter, some people who have no connection with the.... One of the men I talked to, I have a very high regard for, is Professor Panofsky."[79]

Panofsky, upon learning of this exchange, disputed the facts as stated and was asked by Senator Gore if he could testify before his subcommittee on the following Friday. Since he could not leave Washington until Wednesday evening after his testimony to the Joint Committee on Atomic Energy, an aide said to him with a grin, "You got gored by Mr. Gore." He flew to San Francisco Wednesday night and "studied all day Thursday for his testimony" and flew back to Washington on Friday morning to testify "that I did not participate in any advisory capacity to any branch of the Government in reviewing the decision to deploy the current modified Sentinel or Safeguard system. I appreciate having had the opportunity of an informal discussion with Mr. David Packard, Deputy Secretary of Defense, several weeks ago prior to the modified Sentinel decision."[80] Senator Gore inquired, "Was there an extended conversation over a period of time?" Panofsky: "About a half an hour....We happened to accidentally meet at the airport."[81]

In summary, it must be stressed again that invitations for nonadministration scientists to appear before the Senate and House Armed Services and Appropriations committees in opposition to a weapons system were a marked break with tradition. Moreover, a new precedent has now been established and the principle of hearing outside scientific witnesses seems firmly embedded within the congressional mores. One could predict that, just as it once was utterly unthinkable for such appearances, in the future it may well be utterly unthinkable not to have scientists heard in opposition to, as well as in support of, defense and arms control issues.

The Mighty Pen

The scientists' efforts discussed thus far—advising, debating and testifying—were all predominantly verbal activities. In their battles in behalf of or against the ABM, the scientists, not unexpectedly, used their pens as well as their voices. Sixty of the ABM activists produced a spate of newspaper and journal articles, reports, and books. Amongst the earliest scientists to write about ABM was Jerome Wiesner. In discussing comprehensive arms limitation systems in 1960, Wiesner wrote, "It is important to note that a missile deterrent system would be unbalanced by the development of a highly effective anti-missile defense system and, if it appears possible to develop one, the agreements should explicitly prohibit the development and deployment of such systems."[82]

Some of these early ruminations on the ABM were remarkably prescient. Freeman Dyson, who in 1964 opposed deployment of an antimissile system, commented on the intense political pressures existing in both the United States and the Soviet Union to duplicate what the other side does in the strategic weapons areas. Dyson prophesized, "It will require unprecedented self-restraint for the American people to accept a Soviet ABM deployment and not embark on a much bigger deployment in response."[83] Already in 1966, Jeremy Stone pointed out, "We are engaged in one of those debates that takes so long that it changes its form in the course of the dialogue."[84]

The single most cited (whether praised or condemned) treatise on the ABM has been the Garwin-Bethe article that appeared in <u>Scientific American</u> in March 1968. The gestation period for this influential piece was just the proverbial nine months. In June 1967, Hans Bethe was invited to deliver the Mack Memorial Lecture at the University of Wisconsin in Madison. Bethe's talk, which dealt with the ABM, was off the record for the press. However, the presentation was taped and later became the basis for Bethe's discussion of the ABM at the American Association for the Advancement of Science (AAAS) symposium on "Is Defense Against Ballistic Missiles Possible?" held December 26 and 27, 1967.

Gerard Piel, publisher of <u>Scientific American</u>, attended the symposium and suggested that Bethe's talk should become the basis for an article. Bethe and Richard Garwin, another symposium participant, decided to write the article jointly. According to the two authors, the <u>Scientific American</u> article was essentially a "polishing up" of their presentations at the AAAS meeting.

Much confusion exists as to whether the Garwin-Bethe article was submitted to the Department of Defense for security clearance before it appeared in print. According to Bethe, he submitted his AAAS symposium talk to the Defense Department two months before the talk was to be given. Bethe claims he spent the last ten days before the scheduled talk on the phone, urging Defense officials to clear it. Garwin transmitted his contribution to John Foster "for comment, not clearance, and received guidance on questions of classification regarding thermonuclear weapons."

According to Bethe, Foster gave up a Saturday golf date on December 23 to clear the article personally. Although Foster confirmed that he had issued the clearance, a high-ranking army scientist felt that Bethe "had put China years ahead by highlighting specific deficiencies in the U.S. defensive system." A former AEC Commissioner stated that "the Bethe-Garwin article was a shockeroo to me because it divulged so much."

The degree of attention paid to the Garwin-Bethe article was evidenced by the numerous times it was spontaneously mentioned by the interviewed scientists. Typical comments included:

When Garwin says something I really have to convince myself he's wrong if I want to oppose what he's saying.
—a New Jersey physicist

I became convinced that no fundamental overriding issue was being neglected by the opponents when Bethe-Garwin, with access to all the information, opposed it.
—a Harvard professor

The proponents of ABM deployment also took notice of the article's appearance. A memorandum written by Secretary of the Army Stanley Resor to then Secretary of Defense Clark Clifford said that "several highly placed and reputable U.S. scientists have spoken out in print against the Sentinel missile system. Among these are Bethe, Kistiakowsky and Wiesner." To counteract them, Resor proposed,

Although it is difficult because of security aspects to answer the technical arguments used by these men against Sentinel, it is essential that all possible questions raised by these opponents be answered, preferably by non-government scientists. We will be in contact shortly with scientists who are familiar with the Sentinel program and who may see fit to write articles for publications supporting the technical feasibility and operational effectiveness of the Sentinel system. We shall extend to these scientists all practical assistance.[85]

The Resor memorandum was written in September 1968, and the proportion of pro-ABM articles written by scientists increased slightly thereafter. How much of this can be attributed to Defense Department efforts and how much to the general intensification of the debate remains conjecture.

Of the 60 scientists who published articles on the ABM, 39 were Outers and 21 were Inners; 45 were anti-ABM and 15 were pro-ABM. In proportion to their numbers in our total sample, pros and antis were equally prolific with their pens.* What is interesting is the almost universal (and correct) feeling expressed by the pro-ABM writers that they belonged to a small minority. We mentioned earlier that the Acheson-Nitze Committee to Maintain a Prudent Defense Policy was expressly established because of a feeling that the debate was entirely too one-sided. Donald Brennan, writing in Foreign Affairs in April 1969, suggested that "the doubts increasingly expressed in the Congress in early 1969 may well result from the highly one-sided literature on the subject."[86]

In the ABM issue as a whole, both sides generally perceived themselves as "underdogs." Among the writers, however, the feeling was confined exclusively to the pro-ABM scientists. The anti-ABM scientists wrote merrily away and their papers, articles, reports, and books tumbled forth undiminished until the Senate vote in August 1969.

Lobbying

"Lobbying," writes Lester Milbrath, "is an inevitable concomitant of government."[87] By lobbying we mean communication directed to a decision maker with the hope of

---

* Some of the deluge of material written by scientists is accounted for by the fact that often the congressional testimony of the scientists formed the basis for articles reprinted in journals and in books.

influencing his decision. Fully 40% of the ABM scientists lobbied with elected officials in Washington, and 29% lobbied with state and local representatives. There was considerable overlap: 12% of the sample did both.

However, this lobbying was predominantly an activity of the anti-ABMers, as is illustrated in Table 9.*

Many of the pro-ABM scientists appeared to view lobbying by their scientific colleagues with distaste and expressed mild shock at being asked whether they had lobbied themselves. However, it is entirely possible that the very same scientists who seemed to be taken aback by a suggestion of lobbying had in fact maintained contacts with legislators about the ABM which they viewed as merely informational, educational, or social.

It is interesting to note that this seems to a reciprocal phenomenon. Donald Matthews reports that "Senators do not perceive much of the lobbyists' action as lobbying. When asked how much 'pressure' he actually felt, one well-known senator replied, 'You know, that's an amazing thing. I hardly ever see a lobbyist. I don't know, maybe they think I'm a poor target, but I seldom see them.'"[88]

Some of the anti-ABM physicists who attended the American Physical Society meeting in Washington, D.C., in April 1969, engaged in an intensive lobbying effort. The Federation of American Scientists, Council for a Livable World, and the newly formed Scientists and Engineers for Social and Political Action jointly sponsored a "how to lobby" briefing session for the physicists. Between 300 and 350 scientists were instructed in the art of lobbying by Tom Halsted and Jeremy Stone.

Under the leadership of Barry Casper, a physicist at Carleton College in Minnesota, 53 Senators or their aides were seen by the physicists during the three-day meeting. For some physicists, this was their baptism into congressional waters. Reaction to their plunge was mixed, with some scientists expressing surprise at "how responsive the system is to just a little effort." Others became convinced "one needs a powerful polictical base, not just words of sweet reason."

Table 9. Percentage of Scientists Who Lobbied with Elected Officials

	Lobbied in Washington (n = 59) (%)	Lobbied Locally (n = 41) (%)
Anti-ABM	86	98
Pro-ABM	14	2

* This table is based entirely on self-admitted or self-designated lobbying efforts. No attempt was made to ascertain whether a scientist who answered that he had not engaged in any lobbying had in fact spent time trying to convince a legislator of his point of view on deployment.

Not all of the lobbying was done personally in Washington. Jay Orear at Cornell persuaded dozens of his scientific colleagues to write letters to 30 key senators. Orear claims that each of those 30 senators received at least 50 letters and began to remark with surprise that all of their anti-ABM mail seemed to be coming from Ithaca, New York!

These endeavors to exert constituency influence on elected officials were predominantly actions of the anti-ABM group and of Outer scientists. Of the scientists who lobbied in Washington, 68% were Outers. Outers also accounted for 90% of the scientists who lobbied with their local representatives.

It is of interest to look at some of the deviant cases here. Who were the Inners who did lobby in Washington and locally? This is another case in which firmly embedded tradition was shattered during the ABM debate. Historically, members of the President's Science Advisory Committee have considered their advice to the president to be of a privileged nature which must remain confidential. There would seem to be varying degrees of privilege:

1. The content and nature of the president's remarks, requests, or charges to the PSAC,

2. The content and nature of the PSAC deliberations,

3. The content and nature of the PSAC's advice to the president,

4. Individual views of the members of the PSAC.

There seems to be general agreement among past and present members of PSAC about the strict confidentiality of the first three of these classes of information.

In the past, the pervasiveness of executive privilege in effect decreed that PSAC members who disagreed with official policy would not give voice to their opposition. In the ABM case, however, this custom was honored in the breach as well as in the observance. This break with tradition revolved around the fourth class.

Richard Garwin, a member of both PSAC and the Defense Science Board, wrote a personal letter to each senator on July 10, 1969, urging the defeat of the proposed Safeguard deployment. Garwin declared, "Although I personally do not see the need to defend Minuteman at this time, I would support a reorientation of Phase I of Safeguard toward a large number of small radars specifically designed for the task of defending many robust low value targets."[89]

In the spring of 1970, Garwin mailed to senators a Safeguard discussion paper in which he found Safeguard "seriously deficient in the protection which it can give Minuteman if the Soviet threat materializes," and noted that "money in defense of Minuteman would be far better spent on providing real defense from a dedicated system."

Garwin feels that "to support whatever the Administration does, right or wrong, violates the code of conduct for government employees." He suggests that he has an obligation to the government, not only to the executive branch. Therefore, he feels no reluctance in informing senators and congressmen of his own views both on ABM and SST, without disclosing the nature of his advice to the administration.

In summarizing the scientists' lobbying activities, we note that no scientists thought these contacts were instrumental in actually "changing any votes." But many felt: (a) the activities were of educational value to them, gaining them insight into "how the sytem operates"; and (b) it was important that elected officials became aware of the intensity and the extent of the misgivings a sizable segment of the scientific community held about the deployment of Sentinel and/or Safeguard.

Summing Up

In addition to the activities we have been describing, scientists also spoke on radio and television; organized a march to the White House where they presented anti-ABM petitions to Presidential Science Advisor Lee DuBridge; held press conferences leafletted at shopping centers; placed advertisements in national newspapers; arranged "work stoppages" and "work ins"; held rallies and forums; designed and sold buttons, posters, and bumper stickers. Amid this welter of activity, it is necessary to keep in mind the relatively small number of scientists who were thus engaged. The vast majority of the scientific community did not participate publicly in any way.

Those scientists who did become ABM activists contributed considerable amounts of time and energy to the cause. The total time each active scientist spent on ABM activities ranged from less than ten hours to substantial involvement over many years. The modal time was several weeks full time. For the scientists who had been involved with ABM for many years, the main activities were serving on advisory panels, testifying, speaking in other communities, writing, and lobbying in Washington. The major efforts of those scientists, whose involvement with the ABM could be measured in weeks rather than in years, were in speaking in their own communities and lobbying with their state and local officials.

The sector in which a scientist was employed had an influence on the range of his activities, as is shown in Table 10. It appears that industrial employment is less likely to lead to participation in an issue like ABM than employment in the other sectors. The greater freedom allegedly afforded to university professors to partici-pate in the political arena is borne out by the data. The often-stated view of scientists

Table 10.   Percentages of Present Employment of ABM Activists

	Total Sample (n = 152) (%)	Testified (n = 33) (%)	Spoke Locally (n = 76) (%)	Wrote (n = 60) (%)	Lobbied in Washington (n = 59) (%)	Spoke in Other Communities (n = 44) (%)	Lobbied Locally (n = 41) (%)
Academic	60	70	71	64	67	58	71
Industry	16	9	4	12	10	9	—
Nonprofit	7	9	5	9	9	12	—
Government Laboratory	13	3	17	14	9	14	27
Other Government Agencies	3	3	1	2	2	2	—
DOD, DDR&E	2	6	1	—	3	5	2

employed in government laboratories that they work in an academic atmosphere also
seems substantiated by Table 10.

The fervor with which some of the activists participated in the ABM cannot be
captured by statistical relationships or numerical tables. At least two scientists,
one at a government laboratory and one at a college, took a month's leave, unpaid,
in order to be able to devote full time to the ABM fight. Phone bills in the hundreds
of dollars charged to personal accounts were incurred in many instances. One
Argonne scientist who felt he was unable to articulate clearly enough in public forums
took public-speaking lessons. Wives and children, neighbors and friends were enlisted to aid in the battle. To these activists, the ABM issue became an integral
part of their personal lives. They approached it not in the abstract, not with detachment, but with commitment and passion.

THE GOOD GUYS AND THE BAD GUYS

The commitment and passion that scientists brought to the ABM debate were instrumental in transforming an issue into a drama. The participants themselves, if
not the general public, saw in Wohlstetter's vivid words, in his article from which
we borrow the title of this section, "the forces of virtue arrayed against the forces
of evil." Such self-appellations of the good guys and the bad guys, the white-hatted
heroes triumphing over the black-suited villains, would seem comprehensible if the
opponents genuinely differ from each other along a whole spectrum of perceptions,
attitudes, and behaviors.

We are, after all, dealing with a fairly uniform and selected group of highly educated, upper-class, professional men who are usually considered to have a great deal in common with one another. The growing corpus of literature on the sociology of science all implicitly assumes the essential homogeneity of the scientific community and the universality of the mores governing it. Norman W. Storer, in listing seven different approaches by which the community of science has been studied, [90] observes that the ingredient common to all of them is consensus on the "oneness," the homogeneity, and the uniqueness of the scientific community.

Our data suggest that such a view of the community of scientists is an oversimplification. A more appropriate model appears to be one that views scientists as members of the body politic. Their motivations, attitudes, perceptions, and behaviors on a great variety of issues are more accurately viewed as reflections of their political, not their scientific, orientation, consequently exhibiting a range of diversity not corresponding to those attributable to a homogeneous community. In essence, we are asking whether the protagonists differed only in the unidimensional question of the deployment of an antiballistic missile or whether the differences of opinion on the ABM were symptomatic of deeper rifts, fissures, and divisions within the scientific community.

We look first at the background characteristics of the pro- and anti-ABM groups, then at their views and actions on ABM, and at their attitudes on a range of larger issues concerning the scientific community.

Background Characteristics

Age/Length of Involvement with ABM. The opposing sides tended to view each other through caricatured and stereotyped lenses. To some of the proponents of deployment, the anti-ABM scientists were young "whippersnappers," inexperienced in strategic arms consideration, who "had difficulty in separating the signals from the noise." Many opponents of deployment viewed the pro-ABM scientists as "the old guard, defenders of the status quo." How close to reality were these mutual mental images?

With regard to age, first of all, it does appear that the stereotype was based on a real difference. The mean age for the anti-ABM scientists (the "young whippersnappers") was 45.5 years, while for the pro-ABMers ("the old guard") it was 50.2 years. This difference is significant beyond .01. Insofar as age is commonly found to be a critical variable in differentiating political opinion, our finding is in agreement with a large body of relevant research. In a recent and comprehensive national survey of students, faculty, and administrators initiated by the Carnegie Commission

of Higher Education in 1969, Ladd and Lipset found "No other variables discrim-
inate as powerfully among the political opinions of political scientists as those
revolving around age."[91]

Accepting a real difference in age between the two groups, what of experience?
Were the anti-ABM scientists Johnny-come-lately to strategic arms deliberations?
Had the pro-ABM strategists devoted significantly more years to the questions of
defense and deterrence? Here again the data confirm the participants' perceptions.
Of the pro-ABM scientists, 63% indicated that the time they had been involved with
the ABM issue should be measured in years, not hours or weeks, compared to only
23% for the antis.

Scientific Speciality. Several social scientists in recent years have proposed that
political party affiliation and political attitudes of professional disciplines tend to
be significantly influenced by the profession itself and by association with colleagues.
That is to say that professional socialization plays an important role in determining
party affiliation and political ideology. While not entirely consistent with each other,
the findings of these studies can be summarized in Table 11.

Combining the fields represented in our study into the categories of social science,
physical science, and engineering, and arraying them by the percentage of scientists
within that discipline opposing deployment of ABM, we get the results shown in
Table 12.

Table 11. Summary of Findings Regarding Political Views of Professional
Groups

Study	Liberal, Democratic	Moderate	Conservative, Republican
Nichols[92] (1968) (n = 37)	Mathematicians Biologists	Physicists	
Ladd[93] (1969) (n = 18,500)	Political Scientists Physicists	Biologists	Chemists
Turner and Spaulding[94] (1969) (n = 2,389)	Political Scientists Psychologists		Mathematicians Engineers
Eitzen and Maranell[95] (1968) (n = 979)	Psychologists	Chemists Physicists	

Our data are not in agreement with the general finding that social scientists tend to be "more liberal" than physical and biological scientists. However, it must be emphasized that because 67% of the ABM activists were physicists the numbers in the other disciplines are exceedingly small. The differences in Table 12 are marginally significant (P<.05).

Education. We find no significant differences between pro- and anti-ABM scientists in terms of the highest degree received. The figures are as follows: 2% of the anti-ABM and 8% of the pro-ABM scientists report a bachelor's as their highest degree received; those whose highest degree is a master's are 5% of the anti- and 8% of the pro-ABM scientists. All others hold Ph.D.'s, LL.B.'s, or M.D.'s.

Scientific Stature According to the following three criteria of eminence, we find no significant differences in terms of receiving a Nobel Prize, membership in the National Academy of Sciences, or in number of citations listed in the Science Citation Index. The data are presented in Table 13.

Employment. With respect to employment and patterns relating to the funding of research, the picture changes drastically. Table 14 illustrates the relationship between place of employment and stance on ABM.

The highly significant correlation between a scientists' place of employment and his stand on the ABM corresponds to the perceptions of the scientists themselves concerning this relationship. Some 93% of the entire sample answered affirmatively this question, "Do you think one's institutional position influences one's political views and attitudes?" Typical answers were:

Table 12. Percentage of Each Discipline in Sample Opposed to ABM

Physical Science	(n = 119)	77
Engineering	(n = 15)	53
Social Science	(n = 12)	50

Table 13. Scientific Stature of Anti- and Pro-ABM Scientists

	Anti-ABM Scientists	Pro-ABM Scientists
Nobel Laureates	n = 4	n = 2
Members of National Academy of Sciences	n = 20	n = 6
Mean Number Scientific Citations	50	49

Table 14. Place of Employment and Stand on ABM

Employment	Anti-ABM (%)	(P < .001)	Pro-ABM (%)	
Academic	85		15	(n = 89)
Industry	39		61	(n = 23)
Nonprofit	38		62	(n = 8)
Government Laboratory	76		24	(n = 17)
Department of Defense	—		100	(n = 3)
Other Government Agencies	80		20	(n = 5)

Yes, one's outlook tends to be limited by the type of problems one is requested to concentrate on. Thus, in government, a scientist will be asked the technical feasibility of a given weapons system, not if there may be a non-technical solution that might be better.
—University professor

If I were at Harvard and coming up for tenure, it would be harder to come out for ABM than if I were working for Herman Kahn.
—Social scientist

Many scientists did differentiate between attitudes and behavior, pointing out that where one worked would affect how one spent one's time or what one did but not necessarily how one thought or felt about political issues. Others stressed that the correlation between institutional position and political views was not causative but rather a matter of preselection. As a scientist at Los Alamos said,

There is a definite sort of selection process. Those unwilling to consider anything good about weapons will not be here or in industry. They will congregate on the campuses.

Funding Source. A similar finding emerges when we look at the source of funding of the scientists' research. For the entire sample, sources of funding were ascertained for 98 scientists; 35% received support from the Defense Department. Table 15 shows the relationship between having one's research funded by the Defense Department and one's stand on the ABM.

Again the data confirm the spontaneous remarks of many of the scientists. In response to the question: "To what do you attribute the differences between pro- and anti-ABM scientists?" there was almost complete agreement that technical considerations were not paramount. Many of the anti-ABM scientists made statements like

Table 15. Defense Department Funding of Research and Stand on ABM

	Anti-ABM (n = 71) (%)	(P < .001)	Pro-ABM (n = 24) (%)
Funded by Department of Defense	18		79
Not Funded by Department of Defense	82		21

Table 16. Working on Weapons versus Stand on ABM

	(P < .001)(The strength of the relationship is indicated by a gamma value of .78)		
		Anti-ABM (n = 79) (%)	Pro-ABM (n = 20) (%)
"Have you ever worked on weapons development?"	No	72	20
	Yes	28	80

Table 17. Working on Defense-Related Research versus Stand on ABM

	(P < .01 gamma value = .72)		
		Anti-ABM (n = 83) (%)	Pro-ABM (n = 29) (%)
"Have you ever worked on defense related research?"	No	31	7
	Yes	69	93

It very much followed funding lines. If you were funded by DOD, you were for it.
—University professor

The difference was dependence on military funding for support or not.
—University professor

The Hudson Institute were the main people for it and they have support from the Army.
—Former government official

Weapons and Defense Work. Several scientists indicated that they thought whether a scientist had worked on weapons development or other defense-related research was a determinant of his stance on ABM. That this was, in fact, the case is demonstrated in Tables 16 and 17.

Previous Political Activity. We have indicated previously that 72% of the ABM participants had been politically active prior to their entry onto the ABM battlefield. Since the pro-ABM scientists are older and were involved in the ABM issue before the anti-ABM scientists became activists, we could hypothesize that they were also

more active politically in other issues than the anti-ABM scientists. This hypothesis
is not borne out by the data that show no significant differences at all between pro-
and anti-ABM scientists in the amount of time spent on non-ABM political issues.
While this difference did appear in the type of issue these men had been engaged in
(no pro-AMB scientist had been active in antiwar or peace-related activities), the
fluctuations could have been caused by chance alone and were not pursued further.

Views on ABM

Deployment. Before the majority of scientists became active in the ABM debate,
certain key events had already occurred. If scientists approach issues with a common
attitude, as the folklore about them would lead us to believe, then one could assume
that their interpretations of events in the recent past would resemble one another,
allowing perhaps for subtle shadings and nuances to differentiate among them. If
our model of scientists as political men is operative, then we would expect past
events to be viewed through partisan eyes and to fail to cluster neatly together.

On October 16, 1964, the People's Republic of China detonated its first atomic
device, marking China's emergence as a nuclear power. From then on, the threat
of a Chinese missile capability became intertwined with the domestic ABM debate
within the government.

When the deployment decision was announced by Secretary McNamara on September
18, 1967, the official rationale given for the decision was as a "countermeasure to
Communist China's nuclear development . . . (and) to provide an additional indication
to Asians that we intend to deter China from nuclear blackmail . . . and enable us
to add—as a concurrent benefit—a further defense of our Minuteman sites against
Soviet attack."[96]

The scientists in our sample were asked: "What is your explanation of the September 1967 decision to deploy the ABM?" The answers are tabulated in Table 18.

Not a single surveyed scientist, who opposed deployment of the ABM and who
expressed an opinion on the subject, indicated that he believed the official explanation
for the deployment decision. Nearly 60% of those scientists favoring deployment
seemed to accept this explanation. Here, in dramatic form, is our first inkling of
the enormous gulf in the perceptions and attitudes between the ABM protagonists.

On March 14, 1969, President Nixon, in announcing his decision to proceed with
the Safeguard ABM stated that "The Soviet Union has engaged in a build-up of its
strategic forces larger than was envisaged in 1967 . . . ." He continued that "the
Chinese threat against our population, as well as the danger of an accidental attack,
cannot be ignored."[97]

Table 18. Explanation of the Sentinel Decision by Anti- and Pro-ABM Scientists

| Answers | (P < . 001) | |
	Anti-ABM Scientists (n = 52) (%)	Pro-ABM Scientists (n = 22) (%)
Political Pressure	100	27
From Republicans From Congress From the Military		
Growing Chinese Threat	—	59
Growing Soviet Threat	—	14

Table 19. Explanation of the Safeguard Decision by Anti- and Pro-ABM Scientists

| Answers | (P < .001) | |
	Anti-ABM Scientists (n = 48) (%)	Pro-ABM Scientists (n = 19) (%)
Political Pressure From the Cities From Congress From the Military	69	21
"It was a compromise" "Expediency"	29	16
"Growing Soviet threat"	2	53
Needed for SALT Agreement	—	10

Again, the ABM scientists were asked for their explanation of President Nixon's decision and, again, as shown in Table 19, the opposing sides interpreted the events in starkly different ways.

Of the scientists agreeing with the administration's decision, just over half (53%) cited the official rationale as their explanation for the decision, while about one-third (37%) felt the decision was based on the expediency of trying to arrive at a compromise solution, or reaction to political pressure from the cities, Congress, or the military.

Ninety-eight percent of the scientists in our sample who opposed deployment of the Safeguard antiballistic missile defense indicated that the official explanation, the growing Soviet threat, was not, in their opinion, the real reason for the decision.

The numerical base of the evidence is admittedly small, but the strength of the correlations and the unidirectionality of the findings provide a strong empirical foundation for emphasizing the greater explanatory role of the <u>political</u> rather than the scientific milieu for insight into the community of scientists.

<u>Feasibility</u>. Beyond the question of whether scientists believed the administration's reasons for this defensive system, one might assume that considerations of technical feasibility would be of some importance to scientists in determining whether they favored or opposed deployment of such a system. Did scientists who favored deployment believe such a defense was technically feasible? Did scientists who had severe doubts about the technical feasibility of ABM oppose deployment?

Just as views on the respective roles of defensive and offensive weapons in the strategic equation have changed over the decades from the 1950s and 1960s into the 1970s, so have perceptions of the feasibility of various technical solutions to military problems. By the time of the public debate in 1969, there was no dispute regarding the technical feasibility of intercepting a few attacking missiles. The dispute revolved around two other questions.

One concerned the reliability of highly complicated advanced computer technologies under the unprecedented conditions of a sky full of nuclear explosions. Opponents of ABM, such as J. C. R. Licklider, emphasized "the unhappiness that lies ahead for anyone who deploys a large, complex system that involves computers and software, that faces a changing and complicating threat and that cannot be tested continually as a whole."[98]

John Foster tried to counter the "non-testability" argument when he wrote that all individual components (radars, missiles, and computers) have been tested separately and "thus, the only real task that the Safeguard system has, is to integrate all these functions in the computer programs and to check thoroughly and test out the programs before the system is made operational." He went on to say that the simulation would involve simulation tapes, "so that the entire system is exercised just as it would be in a real battle."[99]

This argument was rebutted by Computer Professionals against ABM, an ad hoc organization of 800 members of the computing profession. In a fact sheet prepared in 1971 and distributed to the major news media, Daniel D. McCracken, chairman of the group, stated that "simulating the system, just as it would be in a real battle is impossible because there is no way of knowing exactly what the real battle would be like. And, in any event, simulation is no guarantee that a system will work properly when it is tried out on the real thing."[100]

The second major disagreement concerning the feasibility of an ABM had to do with the adversary nature of an offense-defense exchange. Anti-ABM scientists argued that

1. The technical uncertainties of an ABM would fuel the arms race by causing the attacking country to overestimate and the defending nation to underestimate its effectiveness,

2. The system could always be overwhelmed by sheer numbers; if there were 1,000 interceptors, the 1,001st incoming nuclear weapon would "come in free."

3. The long lead time necessary for deployment of an ABM would enable the offense to devise stratagems to confuse, deceive, or elude the defensive missiles.

Scientific proponents of ABM argued that

1. We will not know whether ABM is technically feasible if we do not go beyond the research and development stage and we cannot afford not to know.

2. Interceptor exhaustion (attacking with more warheads than the defender has interceptors) is not as simple as it appears because the attacker cannot know with certainty the number of interceptors available to the defense.

3. Most of the enormous variety of decoys, chaff, and other penetration aids that have been built and tested by the United States have been ineffective.

4. The proposition that ABM has to work 100% or not work at all is not a standard applied to other weapons. ABM is feasible to some extent and to some effectiveness.

In our own survey, the scientists were asked to differentiate among three possible kinds of antimissile defenses: hard-point, light area, and urban defense. Their assessments as to the technical feasibility of each type are shown in Table 20.

For each type of defense the results in Table 20 are significant beyond .001 and yield gamma values of .98, .98, and .95, respectively.

Our data would seem to indicate that those who opposed and those who supported ABM, each viewed the ABM issue through different sets of judgmental and perceptual lenses. The pro-ABM gladiators wore rosy-tinged spectacles concerning feasibility. Although they saw growing Russian and Chinese threats to our national security, they believed that ABM deployment decisions were taken by two successive administrations to counter those threats and that an antiballistic missile system such as Sentinel or Safeguard was a technically viable and feasible solution to the perceived threats. The anti-ABM battlers looked through the glass darkly. They perceived the deployment decisions as expedient responses to political pressures, as being politically not strategically motivated. Further, they simply did not believe in the technical feasibility of the proposed ABM systems.

Activities. We have already discussed briefly some differences between pro- and anti-ABM scientists in terms of which activities the opponents engaged in, in their quest for a victory on the political battlefield. Altogether 57 ABM scientists served on advisory panels of one kind or another dealing with the question of ABM.

Table 20. Technical Feasibility of HARD-POINT, Light Area, and Urban Defense
Judged by Anti- and Pro-ABM Scientists

Type of Defense	% Answering "It is feasible" or "probably feasible"	
	Anti-ABM	Pro-ABM
	(n = 83)	(n = 36)
Hard-Point Defense	43	97
Light Area Defense	21	97
Urban Defense	1	52

Fifty-two percent of these were pro- and 48% were anti-ABM. When we remember
that overall our sample opposed ABM 73% to 27%, we must conclude that propor-
tionately twice as many pro-ABM scientists served on these committees as would
be expected if they were randomly drawn from our sample.

We can interpret this finding in several ways: one line of reasoning would be
that scientists who served on advisory panels become the most knowledgeable about
the subject of antiballistic missile defense, and, having become informed, they
recognized its virtues.

An alternate explanation is that the scientists on advisory panels were chosen not
solely for their technical expertise but also for their presumed or known stance on
the desirability of deployment. From the evidence we have presented concerning
the previous employment and research funding patterns of scientists who serve on
PSAC and DOD panels, this interpretation of scientists serving as "legitimizers"
of the sponsoring agency's basic viewpoint on the deployment of ABM cannot be
·discarded.

One could also conjecture that service on a scientific advisory panel tends to inhibit the expression of antiadministration views. That is to say, scientists become
co-opted to the adminstration point of view. If one accepts this notion, it would not
be surprising to find a majority of advisory panelists favorable to ABM. However,
the persistent, nearly unanimous opposition of PSAC and PSAC panels to ABM for
more than ten years and spanning three different administrations would seem to
invalidate this argument.

Whichever interpretation one gives credence to, what is significant is the fact
that of the 40 pro-ABM scientists in our sample, fully 65% served on DOD or PSAC
panels, while among the 106 anti-ABM scientists, only 23% had done so.

Quite a different ratio is observed when we turn our attention to lecturing or debating within one's own community. Of the 76 scientists who spoke publicly to their

friends and neighbors on the subject of ABM, 12 were pro-ABM and 64 were anti-ABM. Two background characteristics may serve as partial explanations for this finding: (1) The anti-ABM scientists were, on the average, the younger men. Addressing town fathers and citizens, fraternal organizations, and political bodies, night after night, lunch hour after lunch hour, requires stamina and endurance—as well as naïveté and foolhardiness, according to some—that one more likely expects to find in younger men. (2) Nearly three-quarters of the anti-ABM scientists were academics, compared to only one-seventh of the pro-ABMers. Traditionally, university and college professors have felt free and, indeed, often obligated to participate in public discussions of current and controversial issues. As an M.I.T. physics professor bluntly stated, "It's a lot easier and cheaper for academics to stand up and be counted in issues like ABM."

Many industrial and government scientists do not feel that they possess the same freedom and security to become publicly identified in a controversial issue. Whether, in fact, that is illusion or reality is immaterial. It is the scientists' perceptions of constraints upon their outside-work activities which serve as inhibitors upon their participation in issues like ABM.

When we consider those scientists who traveled to other communities to discuss or debate the virtues or vices of ABM deployment, the differences between pro- and anti-ABMers (which were significant beyond .001 in the case of local speaking engagement) vanish completely. Of the travelers, 30% of the anti- and 33% of the pro-ABM scientists lectured or debated in communities other than their own.

To account for this finding, we need to remember that pro- and anti-ABMers did not differ significantly in their eminence or scientific stature, as was shown in Table 13. In addition, an invitation to appear before a community in another locale, particularly one where no missile site was to be located (and where no "protest" might thus be expected), is one that pro-ABM scientists would by inclination be much more likely to accept. We have already mentioned the distaste with which many pro-ABM scientists viewed lobbying. Addressing local town fathers who were about to vote on the siting of missiles in their vicinity is more likely to be construed as a form of lobbying than is journeying afar to deliver the same speech.

When we examine the testifying activities of our ABM scientists before congressional committees, we also find no significant differentiation between pro- and anti-ABM appearances: of the 32 scientists who testified, 20 were opposed to deployment and 12 were in favor of deployment. Most committees, in fact, made a careful effort to have both sides represented and scheduled the scientists in prudently balanced pairs or groups.

The first scientists to appear before the Gore subcommittee on March 6, 1969, were Daniel Fink, a proponent of deployment, and J. P. Ruina and Hans Bethe, opponents. The Senate Armed Services Committee on April 22, 1969, paired Paul Nitze and William G. McMillan against Herbert York and Wolfgang Panofsky. Two days later, the protagonists before the House Committee on Armed Services were Marvin Kalkstein, George Rathjens, and Frank Collins in the opponents' corner and John Wheeler, Donald Brennan, and Lawrence O'Neill in the proponents' corner.

The Gore subcommittee in 1970 heard only anti-ABM scientists as nonadministration witnesses although pro-ABM scientists were invited to testify and declined to do so. Secretary of Defense Melvin Laird and John Foster testified on the administration's behalf. The one-sidedness of these particular hearings was frequently mentioned by pro-ABM scientists—never however by ABM opponents!

In the realm of publications, we find no significant differences between the number of articles published by the pro's and the anti's in proportion to their numbers. However, in absolute terms, only 15 pro-ABM scientists had articles appear in the public press compared to 45 anti-ABMers. Explanations for this difference varied according to one's stance on ABM. Pro-ABM scientists claimed that they could not get their articles published due to bias on the part of the media. The anti's were inclined to feel that the amounts published by each side were fairly accurate reflections of the division of sentiment within the scientific community on the ABM issue.

From Table 11 we learned that lobbying in Washington or with state and local officials was primarily an activity reported by anti-ABM scientists. Only 20% of the pro-ABM group lobbied in Washington and only 3% lobbied locally, compared to 49% and 40%, respectively, for anti-ABM activists. Again, the sector of employment must be considered as a critical variable. With cries about the military-industrial complex echoing across the country, scientists employed by industry or by the government undoubtedly felt constrained in their ability to lobby for or against the ABM. The largest ABM contractor, the Bell Telephone System, has for many years had a firm company policy of "avoiding public discussion of the technical facts about ABM systems, even if this information is declassified or has been made public by others, because such discussion might be construed as an attempt to influence national opinion."[101]

Views Concerning the Debate

Thus far we have compared the ABM gladiators by their background characteristics, by their judgments of the reasons for deployment decisions, by their assessments

of the technical feasibility of antiballistic missile defenses, and by the political endeavors they engaged in. In all but the background characteristics, we have found
significant differences. This has been taken to give credence to our view of scientists as members, not of a homogeneous scientific community, but as members of
a heterogeneous political community. Now we shall turn to the question of how these
participants viewed the ABM debate itself.

Did they think the debate had any effect upon the scientific community? If so,
did they perceive it as beneficial or detrimental? How did they see the scientific
community divided on the deployment question? How important did each side think
access to classified information was? How did the opposing sides view each other?
How effective or influential were the scientist-participants seen by each other?
Effect upon the Scientific Community. The large majority of the scientists felt that
the debate did indeed have consequences for the scientific community as a whole.
Of the 116 scientists with whom this question was discussed, only 17% of the anti's
and 12% of the pro's answered that the ABM had little or no effect upon the scientific
community. (See Table 21.)

Of the 98 scientists who felt the ABM issue did have repercussions, 34 felt these
side effects were harmful to the community of scientists. The most frequently given
answer of scientists holding this view was that the debate had harmed the prestige of
the scientific community. Scientists expressing this point of view were rather evenly
divided between the two camps: 14 were anti- and 20 were pro-ABM. The second
most frequently mentioned harmful consequence of the ABM debate was that it contributed to the decline in funding for science. This linkage between the ABM controversy and funding cutbacks in science was mentioned by nine pro-ABM and three
anti-ABM scientists. Also mentioned as a detrimental result was that the ABM debate
produced deep rifts and schisms within the scientific community. The perception
of the scientific community split asunder by the ABM issue was held by pro-ABM
scientists by a 4-1 margin over anti-ABM scientists. Almost 10% of the sample felt
that the debate had produced mixed results; these were usually described as harmful
in the short run and beneficial in the long run. The ABM debate was seen as having
had positive and beneficial effects upon the scientific community by 46% of the sample.
As can be seen in Table 21, these were overwhelmingly anti-ABM scientists.

The division of the ABM participants on the question of what effect the debate had
upon the greater scientific community is summarized in Table 21. From this table
it would appear that the combantants had switched glasses: the pro-ABM scientists
now view things darkly, over two-thirds of them felt that the ABM debate had produced detrimental side effects for the scientific community. The anti-ABMers donned

Table 21. Effect of the ABM Debate upon the Scientific Community as Seen by
Anti- and Pro-ABM Scientists

	Anti-ABM Scientists (n = 83) (%)	Pro-ABM Scientists (n = 33) (%)
	(P < .001)	
The Debate Had No Effect	17	12
The Debate Had Harmful Effects	13	70
The Debate Had Mixed Effects	11	6
The Debate Had Favorable Effects	59	12

Table 22. Division of the Scientific Community on the ABM Issue as Seen by
Anti- and Pro-ABM Scientists

		Majority Oppose (%)	Even (%)	Majority Favorable (%)	
Academic	Anti-ABM scientists	99	—	1	(n = 108)
	Pro-ABM scientists	85	7	7	
Industrial	Anti-ABM scientists	67	16	17	(n = 93)
	Pro-ABM scientists	33	29	38	
Government	Anti-ABM scientists	83	6	11	(n = 76)
	Pro-ABM scientists	32	23	45	
Knowledgeable	Anti-ABM scientists	85	8	7	(n = 47)
	Pro-ABM scientists	24	24	52	

the rose-colored lenses, and over 75% of them perceived the fellowship of scientists
either as unaffected by the ABM controversy or as having benefited from the debate.
Division of Scientific Community. The question of how each side saw the scientific
community divided on the issue of ABM deployment was posed in the following
manner: "How would you estimate the division of the scientific community on the ABM
issue: 90-10 opposed; 75-25 opposed; 50-50; 75-25 favorable; 90-10 favorable?" In
Table 22, these categories are combined to form the simplified groupings of "a
majority opposed," "even," and "a majority favorable."

The respondents were asked about four component parts of the "scientific community": academic, industrial, and government scientists, and scientists knowl-

edgeable (as against those uninformed) on the ABM issue. Table 22 shows the com-
posite view.

Table 22 may be interpreted in the following sense: 99% of the anti-ABM scientists, regardless of where they themselves were employed, viewed the academic
community as opposed to deployment, and only 1% thought that a majority of the
academic community favored deployment. Of the pro-ABM scientists, again,
regardless of where they themselves were employed, 85% also saw the academic
community as opposed to deployment, 7% thought college and university professors
were evenly divided on the subject, and 7% felt that a majority of the academic
community favored deployment. The academic community was viewed as overwhelmingly opposed to ABM by both sides.

The proponents of ABM saw the industrial scientific community as rather evenly
divided among the three categories, while anti-ABM respondents felt that over two-
thirds of the industrial scientists opposed ABM. The sharpest divergences in the
perceptions of the division of the scientific community on the ABM arose in the
case of government scientists and scientists judged knowledgeable about the ABM
issue by their peers. In both these cases, the anti-ABM gladiators saw the vast
majority of their colleagues as arrayed with them, pro-ABM scientists perceived
only a minority of each category opposing ABM.

The proponents of deployment were in their own eyes, clearly holders of a minority
viewpoint within the scientific community. As with the other aspects of the debate
we have examined, the scientists' perceptions of the division of the scientific community were correlated with their own stance on the issue. *

Access to Classified Information. The scientists varied greatly in their perceptions
of the importance that access to classified information played in the debate. A
sizable minority (44%) of the ABM participants stated that they did not think classified
information was very important in the ABM issue. (Of the 52 scientists who expressed
this view, 44 were anti- and 8 were pro-ABM.) Two very prominent scientists, one
a former proponent and a member of the General Advisory Committee of the Arms
Control and Disarmament Agency and the other, a former PSAC member, voiced
almost identical opinions that there was nothing they had learned from their highly
classified briefings that they could not have obtained from reading the New York
Times or Aviation Week and Space Technology. The PSAC scientist recounted that
once after attending a highly classified briefing he asked the briefing officer, "What

---

* Except for the case of the industrial scientists, the differences in the judgments
between the protagonists were significant beyond . 01, with gamma values of . 85,
.76, and .85.

is so secret about all of this? There is nothing that you told us that has not already appeared in print." "Ah," replied the government official, "but now you know it's true!"

The majority of ABM activists, however, did feel that access to classified documents had played an important part in the ABM debate. A famous West Coast physicist, who had spoken out frequently against classification and secrecy in government, felt that "the great mass of scientists were talking of things they did not understand because of secrecy." He asserted that "the distorted statements by opponents of ABM could not be answered because of existing security regulations." Several ABM proponents felt that "opponents of ABM were much more daring in deciding what was unclassified." "I wouldn't have felt so free," said Charles Herzfeld.

More scientists were of the opinion that the issue of access to classified infor- mation and the access itself had benefited the cause of ABM opponents than felt it had benefited the proponents' side. "The opponents could speak with confidence about the performance of the system because of their knowledge."

Scientists who took the opposing point of view, namely that the classified infor- mation issue aided the ABM proponents, declared that "the government made maxi- mum use of disguising certain things by selective declassification" (a government laboratory scientist); and "proponents of deployment in the Senate still vote on the basis that the Pentagon must know what it is doing" (a physicist at a nonprofit institution).

Table 23 summarizes the views expressed by the ABM scientists on the issue of the importance of access to classified information in the ABM debate. Once again, the ABM participants discerned the same subject, the importance of access to classified information, from entirely different perspectives and came up with op- posing views. Pro-ABM scientists perceived access to classified information as much more important than did the anti-ABM scientists, and a majority felt it benefited both sides equally.

Perceptions of Opponents. Since the impression of the ABM gladiators on so many aspects of the ABM debate seem dependent on their own stance on the ABM issue, we would expect this to carry over into their perceptions of each other. We have seen that pro-ABM scientists believed (1) that the ABM debate was injurious to the scientific community; (2) that scientists were sharply divided on the issue; and (3) that the anti-ABM activists were "young whippersnappers" who did not have the experience in strategic matters that qualified them to make pronouncements on the issue. We might, therefore, hypothesize that they would view their opponents as

Table 23. Role of Access to Classified Information in ABM Debate as Seen by
Anti- and Pro-ABM Scientists

	(P < .005)	
	Anti-ABM Scientists (n = 84) (%)	Pro-ABM Scientists (n = 33) (%)
It Was Not Very Important	52	24
It Was Important and Aided Mostly ABM Opponents	16	12
It Was Important and Aided Mostly ABM Proponents	12	9
It Was Important but Gave Equal Advantages	19	55

Table 24. Objectivity of Scientists in the ABM Issue as Viewed by Anti- and
Pro-ABM Scientists

	(P < .001; gamma value = .72)	
	Anti-ABM Scientists (n = 95) (%)	Pro-ABM Scientists (n = 39) (%)
Yes, They Were Objective	24	—
Some Were, Some Were Not	19	8
The Other Side Was Not	17	56
No, They Were Not	40	36

(1) emotional rather than objective; (2) unknowledgeable on the subject of ABM; and
(3) ineffective in the political arena.

Anti-ABM scientists professed (1) that the ABM debate was generally beneficial
for the scientific community; (2) that the great majority of scientists shared their
opposition to ABM deployment; and (3) that their opponents were members of the
"old guard." For them, we would hypothesize that their view of their opponents
would be as follows: (1) emotional rather than objective; (2) knowledgeable on the
subject of ABM; and (3) effective in the political arena.

No norm is considered to be more basic to the scientific ethos than the norm of
objectivity—"the suspension of judgment until 'all the facts are in' and a detached
scrutiny of beliefs in terms of empirical and logical criteria."[102] The ABM partic-

ipants were asked during the interview whether, in their opinion, scientists in the
ABM debate had followed the scientific canons of objectivity. The data are tabulated
in Table 24.

The "tongue-in-cheek" remark of one West Coast physicist was representative
of a large number of respondents who expressed the view that some scientists were
and some were not objective:

I can't help but feel that those who agreed with me were "scientific" and the others
were not.

The opinion that "the other side" was not objective was expressed frequently by
pro-ABM scientists. Two industrial scientists phrased it in these terms:

I think anti-ABM scientists were less objective than proponents. A number of people
against the ABM took the feeling that the country is spending too much on defense,
that the Vietnam war is a bad thing, that the military-industrial complex is pushing
the country . . . . ABM was one issue on which they could get the country and
Congress to focus . . . . It got to be an emotional issue. They let their technical
judgments be colored.

I'm incensed at the dishonesty in the positions of the opponents. They represent
themselves as experts. They clearly are not.

This was echoed by a Nobel Laureate opponent of deployment:

I was anti-ABM; most of my colleagues on this side knew much less and were less
ready to deal "objectively" with the problem. I think their reactions were mobilized
by some crowd-fever and by a generalized anti-administration and anti-DOD
stance . . . .I do not identify many scientists on the pro-ABM side. Perhaps because
they were so isolated I have the impression that many of their remarks were more
cautious. But, after all, ABM is not primarily a scientific issue.

The best-known spokesman of those who contended that neither side displayed
"scientific objectivity" in the political arena on the ABM issue is Albert Wohlstetter, who has written on this subject since 1963. Wohlstetter feels that "when
a scientist represents himself as a scientist, as an expert, as having arrived at
his beliefs as a result of careful analysis of the evidence," he has great obligations.
In Wohlstetter's opinion, these obligations were often not met during the ABM debate.
Wohlstetter states that the scientists'

reputation for professional competence is not sufficiently affected by statements made
out of their field appearing in non-professional journals and never, or very seldom,
subjected to professional scrutiny in professional journals. They might behave more
responsibly if such scrutiny were normal. [103]

The first of the three hypotheses for each side would seem to be confirmed: a majority of the ABM scientists on both sides viewed the behavior of their colleagues in the
political arena as emotional rather than objective. The ABM advocates were much

more inclined to attribute this lack of objectivity to "the other side" than were the ABM opponents.

On the subject of opponents' knowledgeability, 31% of the respondents questioned thought their scientific opponents were poorly informed or uninformed on the ABM issue. Of these 36 scientists, 15 were ABM opponents and 21 were ABM proponents. Twenty-seven percent of the respondents felt the scientist-participants were well informed; these were divided 27 anti- and 4 pro-ABM. The remainder answered that "some of the participants were informed, some were not"; "they were informed on the technical aspects but drew the wrong conclusions"; and "they were knowledgeable about narrow parts of the problem, only." Thus, the data confirm the hypothesis that the pro-ABM scientists cast their scientific opponents in a considerably more negative light than did the anti-ABM scientists.

On the question of how effective and influential each side felt the other had been in the debate, our prediction was that the anti-ABM scientists would attribute greater influence to the scientific community in the ABM issue than would the ABM proponents. The data, as presented in Table 25, support the prediction.

To summarize the perceptions each side had concerning the ABM debate, we found that ABM proponents believed the debate to have been detrimental to the scientific community. They stated that the controversy had harmed the prestige enjoyed by the scientists; it had contributed to the funding crises that scientists face; and, it had brought divisiveness to the collegiality of scientists.

ABM advocates saw the scientific community rather evenly divided over the desirability of ABM deployment. From their perspective, access to classified information was an important ingredient in the ABM issue. They saw their opponents as uninformed and lacking in objectivity and attributed less influence overall to the scientific community in the ABM controversy.

By contrast, ABM opponents thought the debate had benefited the greater scientific community; they saw a vast majority of their colleagues sharing their opposition to deployment. They did not attribute so much importance to having access to classified information; they were more likely to grant that their opponents were objective and knowledgeable; and they felt that scientists had played an influential role in the ABM drama. The model of a monolithic scientific community coolly assessing a given set of data and, after due deliberation, arriving at consensus, if not unanimity, would appear to be superseded by Oliver Wendell Holmes' observation, "All I mean by truth is what I can't help thinking."

Views about Science. Our findings on perceptions of the ABM debate suggest that scientists from the opposing sides might also differ in their views on broader issues

Table 25. Effectiveness of Scientists in the ABM Issue as Viewed by Anti- and
Pro-ABM Scientists

	Anti-ABM (P < .005) Scientists (n = 104) (%)	Pro-ABM Scientists (n = 36) (%)
Scientists Had No Influence	4	22
Scientists Had a Little Influence	5	14
Scientists Had Some Influence	13	8
Scientists Were Influential	79	56

relating to science. Do the ABM scientists who viewed the debate through such
disparate lenses also wear different conceptual and perceptual spectacles when
they look at the profession of science to which they belong? How do they view, not
the objectivity of their opponents, but their own objectivity, both in pursuing scientific research and in their political goals regarding deployment of a defensive system?
How do they feel about the politicization of professional organizations and what do
they see as the proper political role of scientists and of scientist advisors?

Objectivity. Two questions attempted to deal with the scientist's feelings about the
norm of objectivity, presumed to be intrinsic to the ethos of science. First, we asked,
"Did you perceive any tension or conflict between your role as a scientist and your
role in the political arena?" The prolix responses ranged along a spectrum from
"no" to "yes" with many qualifiers en route. Starting on the negative end, typical
answers were

No, because I am conscious of preparing my own rationale. My emotions tell me
in what direction I want to go. I have to see the scientific reasons for getting there.
—University physicist

No tension, but a continuous responsibility to make clear to listeners which of my
judgments were based on professional knowledge and which were my opinions as a
citizen.
—University physicist

Some scientists felt the only conflict was that of time. The time they spent on the
ABM issue was time away from their research and that produced a certain amount
of tension for them.

Several described conflict arising from interpersonal relationships. A pro-ABM
social scientist said sorrowfully,

Some find it odd that being pro-ABM can be consistent with a liberal view of politics and arms control and arms reduction. I am not an arms racer . . . . I felt uncomfortable with my bedfellows, like Senator Thurmond and Representative Rivers. It was unfortunate that the debate polarized views in this way so that to be pro-ABM was to be rightist.

A physicist in Michigan stated that

the conflict is not conflict in principle. It comes from the ethics regarding facts between me and people who are professional organizers and spokesmen for political groups. The particular frustration which I felt was that they were less concerned that the ABM be defeated than that the ABM should be used somehow to radicalize the people.

The second question probed whether the scientist felt he personally had been objective when he dealt with the ABM issue. "Did you state your assumptions explicitly and withold your judgment until all the data were in?" Here, many respondents adopted a fatherly attitude toward the interviewer. "You'd be surprised how nonobjective physics is," said Francis Low. According to Franklin Long, "We aren't very bloody objective in our scientific research either." From Paul Doty, "In the case where half a dozen people are working on the same scientific problem, there also are compromises with scientific methodology."

ABM proponents were much less likely to concede that, in scientific research, objectivity might not be omnipresent. For example, a "think tank" scientist said, "I just solve problems—I try—I don't always succeed. It is possible to detach oneself from the political issue and make a technical problem out of it. If the President wants an area defense, I ask what is the best technical answer."

Table 26 presents the data concerning the scientists' view of their own objectivity in the political arena on the ABM issue.*

Politicization of Professional Organizations. The ABM debate occurred during a time of considerable ferment within professional scientific organizations. The debate over the "Schwartz Amendment" to the American Physical Society Constitution in 1968;[†] the formation of SESPA (Scientists and Engineers for Social and Political Action) and of Computer Professionals Against ABM in 1969; and the political actions by radical scientists at professional meetings of the American Microbiological Society, the Association for Computer Machinery, and the American Association for the

---

*The relatively even split among ABM opponents was examined further to determine if age or scientific stature served as intervening variables. Neither did.

[†]The Schwartz Amendment, introduced by Charles Schwartz, professor of physics at the University of California at Berkeley in 1967, would have amended the Constitution of the American Physical Society to permit the Society to take a stand on political issues such as the Vietnam War. The amendment was defeated by a nearly 3 to 1 vote in 1968. However, since then, the American Physical Society has begun formal reconsideration of its political role.

Table 26. Objectivity in the ABM Controversy as Viewed by Anti- and
Pro-ABM Scientists

	Anti-ABM (P < .01) Scientists (n = 87) (%)	Pro-ABM Scientists (n = 30) (%)
"Were You Objective?"		
No	54	23
Yes	46	77

Advancement of Science throughout 1970, all combined to make the issue of politicization of professional organizations a timely one to discuss with the ABM participants.

The respondents were asked whether they discerned any changes within the professional organizations to which they belonged. Were such organizations, in fact,
becoming politicized? A large majority of both pro- and anti-ABM scientists
answered affirmatively (67% of the anti's and 80% of the pro's). Then, the scientists
were asked to discuss how they felt about such politicization. Did they view it with
approbation or with apprehension? One-third of the scientists voiced strong disapproval of any politicization of professional societies. Typical comments were:

Politicization of professional people is a good thing. Politicization of professional
organizations is probably bad. It reduces our credibility and justly so, when public
statements are made on the technical side of arbitrary issues.
—Anti-ABM graduate student

The Federation of American Scientists exists and is the proper vehicle for that job.
The American Physical Society is tax exempt. It is clear that the people who want
to politicize it are a small group now. There is not the slightest chance they would
win any kind of vote; they would get even fewer votes now.
—Pro-ABM physicist

Just less than half of the respondents (49%) took a moderate or middle of the road
stand. They discussed both pros and cons and on balance favored some politicization.

I think more exchange of viewpoints on controversial subjects would be enlightening
to members and make meetings more worthwhile.
—University physicist

It's a tricky business. I would like to see professional organizations involved with
social, not political issues. It's not hard to say that the American Physical Society
(APS) should be involved with the uses of the physicists' works. I would oppose having
the APS vote on Vietnam but would favor it coming out with a statement on disarmament.
—University physicist

Less than one-fifth (18%) of the sample favored outright politicization.

I see no particular reason why societies of physicists would be more virginal than the AMA or the NAM. They all have developed political action arms.
—University physicist

If physics wants to survive, it better get political.
—Physicist at nonprofit institution

Every organization is formed because some need has arisen demanding organization. Every organization has to change. The APS is a society of physicists and physicists have to relate to things. The society that isn't relevant will fall apart.
—Government laboratory scientist

The data are summarized in Table 27.

Political Role of Scientists. Our respondents were, with few exceptions, scientists who had entered the political arena in order to aid or oppose the deployment of a weapons system. Since they themselves had forsaken their reputed ivory towers and had engaged in the hurly-burly of politics, the question arose as to how these men viewed the proper political role of a scientist? Do they agree with Robert C. Wood's contention that the scientific community is

an apolitical elite, triumphing in the political arena to the extent to which it disavows political objectives and refuses to behave according to conventional political practice?[104]

Or is Wallace Sayre's dictum that

Scientists are now inescapably committed to politics if they hope to exercise influence in the shaping of public policy, including science politics . . . they are effective in the degree to which they understand the political process, accept its rules and play their part in the process with more candor than piety, accepting gladly the fact that they are in the battle rather than above it.[105]

more congruent with their sentiments?

At one end of the range of opinions expressed, were the feelings that scientists best should eschew politics. According to a prominent scientist, "The proper role of scientists is to make science. Our job in science is (1) to do science, (2) to apply science, (3) to explain publicly what has been accomplished. Then, a scientist has no more responsibility regarding the political consequences of his work than a general or a legislator." Only seven scientists in our sample agreed that scientists should be essentially apolitical. Of these, six were pro- and one was anti-ABM.

Moving along the spectrum, one encounters those scientists who stressed the obligation of scientists to make sure that technical considerations are factored into the decision-making system. Some of these scientists stressed the importance of the "insiders" role. Said a scientist at Los Alamos, "Coming out in the open weakens one's effectiveness in the inside world." An independent consultant observed that

Table 27. Desirability of Politicization of Professional Organizations as Viewed by Anti- and Pro-ABM Scientists

| | (P < .001; gamma value = -.89) | |
	Anti-ABM Scientists (n = 75) (%)	Pro-ABM Scientists (n = 21) (%)
"Should Professional Organizations Be Politicized?"		
No	20	81
Moderately	57	19
Yes	23	—

a scientist like himself "feels he loses his effectiveness if his name gets in the paper too often, his is an insider role."

Next come those scientists who professed that scientists should have just the same political role as any other members of the citizenry. As Eugene Wigner phrased it,

Scientists have no more right to decide in which way their contributions to society should be used than have the other members of society.

From John Foster,

It bothers me that some scientists should feel they have a voice beyond ordinary people.

Other scientists agreed with the need for scientists to inform but felt strongly this should be done publicly. A government scientist declared,

Every professional group has a societal obligation. A doctor who sees child abuse has an obligation to report it, to bring it to the public's attention. The scientist has the same responsibility.

According to a university professor,

The model of the ABM is the appropriate model. Scientists acted just right. They got involved; they educated the public and the decision makers.

Working through and with existing organizations was deemed most suitable by still other respondents. They were almost evenly divided between those advocating that the scientists' efforts should go to scientific organizations like the FAS and Council for Livable World and those who thought the political parties were more suitable.

The position most diametrically opposed to the apolitical stance advocated by some was taken by others who feel scientists should assume an advocacy position in the political arena. These respondents opined that the scientists referred to by Wood's "apolitical" appellation abstained from politics because they did not need to lobby. "Physical scientists were the darlings of government. They didn't want to kill the goose that laid the plutonium egg" was how a Lawrence Radiation Laboratory physicist phrased it. According to Charles Schwartz, "It is wrong to give away one's birthright as an opinionated person." Respondents who favored this more activist stance for scientists also stated that "politicians should take appropriately skeptical views of what scientists have to say."

The division between pro- and anti-ABM scientists on the question of the proper political role for scientists is shown in Table 28.

Political Role of Scientist Advisors. The preceding question dealt with the generalized role of scientists in the political arena. The more specific issue of the role of a scientist who has been asked to serve in an advisory capacity to government was also explored. Here, as elsewhere, the lines were sharply drawn. Compare the views of Donald Hornig, former science advisor to President Johnson:

By remaining within the system the leverage in some respects is much greater, but there are corresponding restraints. My conclusion is that some people should certainly participate within the system as the Strategic Military Panel did and I did. For one thing, pressure could be brought and arguments adduced to abandon the whole ABM idea. That was done and we failed, but I doubt that the same people could have done any better on the outside. In the course of it, not only did the system get trimmed down (I would hate to claim as a result of whose efforts), but quite a lot of much better thinking about ABM in the nuclear strategic context was gradually implanted in other thinking. I think there was a considerable net gain, although the big battle was lost.

with those of Owen Chamberlain, Nobel Laureate:

I've reached the conclusion those IDA (Institute for Defense Analysis) scientists are being used. They're not listened to. The best way to stop being used is to drop out.

And those of Lewis Branscomb, former director of the National Bureau of Standards:

While scientists like Professor Panofsky have been meticulous in their protection of public trust that exposure to classified information entails, how could the minority of senators have had a way to get the information if Panofsky and others hadn't served in the IDA and PSAC?

with Gordon Kane, professor of physics, University of Michigan:

The members of the executive branch science advisory apparatus clearly took a strong political stand by allowing the government to claim that studies showed that it would "work." They should have spoken up publicly or at the very least dissociated themselves from the executive's claims. The role they played was thus of a "retained lawyer" rather than that of a "scientific advisor."

Table 28. The Proper Political Role for Scientists as Viewed by Anti- and Pro-ABM Scientists

| Scientists Should | (P < .001; gamma value = -.81) | |
	Anti-ABM Scientists (n = 65) (%)	Pro-ABM Scientists (n = 28) (%)
Be Apolitical	3	21
Inform Policy Makers	9	54
Act as Any Citizen	25	14
Inform the Public	25	7
Work through Existing Political Channels	23	4
Play an Advocate's Role	15	—

Finally, those of Stanislaw Ulam, professor of mathematics, University of Colorado:

It is good for scientists to be involved a little in defense; otherwise charlatans will do it. Although in several respects I do not agree with the views and policies of the present administration, if I were asked for advice, I would give it.

with Gregory Dash, professor of physics at the University of Washington:

Inside advisors are prostitutes.

The viewpoints of scientists who felt changes were needed in the science advisory apparatus, that scientist advisors should take more of an advocate's stand, should not accept constraints, and should see their role as advisors to a larger constituency than just to the executive department have been grouped together in Table 29 under the heading "Changes Needed, More Activist." The body of responses indicating general acceptance of the present role of scientist advisors is labeled "Acceptance of Existing Patterns." The division between pro- and anti-ABM scientists is as shown in the table.

The picture that has emerged from our examination of the ABM scientists' views on science is one of a multifractionated, hetereogeneous community, whose members exhibit sharply divergent cognitions, attitudes, and perceptions.

In broad brushstrokes, we can describe pro-ABM scientists as having internalized the norm of objectivity, and the norm has become a perceptual screen through which they view the science-politics interface. Since in their view anti-ABMers were not objective and they themselves were, it became easy to denigrate and castigate the

Table 29. Role of Scientist Advisors as Viewed by Anti- and Pro-ABM Scientists

	$(P < .001$; gamma value $= .95)$	
	Anti-ABM Scientists (n = 49) (%)	Pro-ABM Scientists (n = 30) (%)
Changes Needed, More Activist	76	7
Acceptance of Existing Patterns	24	93

opponents for forsaking the "scientific" path in the ABM controversy. Pro-ABM scientists voiced general disapproval of politicizing professional organizations and of having scientists and scientist advisors assume more public, more Ralph Nader-like political stances. Since such changes would alter the "objective" image of science, these views are self-consistent.

Anti-ABM scientists were more willing to accept the fact that in the political arena they are not more chaste than any other group. Accordingly, they more often expressed tension caused by this role conflict. If one believes that pure objectivity is more myth than reality in the actual practice of science, then one is more likely to tolerate, if not accept, a more activist stance by scientific organizations and by scientists within and outside the scientific organizations and by scientists within and outside the science advisory network. This was the prevailing viewpoint of anti-ABM scientists.

INFLUENCE IS IN THE EYE OF THE BEHOLDER

The fundamental question that remains is what effect this flurry of activity had upon the actual events as they occurred over time. First, as to the deployment decision of September 1967: Were scientists involved in the making of that decision? What role did scientists play in the initial site selections and in the later changing of several of these? To what extent were scientists responsible for the change in emphasis from a light area defense to a defense of Minuteman bases and/or the National Command Centers and the consequent removal of the sites from large metropolitan areas? How much credit or blame should be apportioned to these scientists for the historic 50-50 tie vote in the Senate?

To deal with a general evaluation of the influence of the scientists we need to differentiate between the scientists' own perceptions of their effectiveness and assessments by the decision makers themselves. Likewise, we must distinguish the arenas where scientific influence can be measured: in the executive and legislative branches, and in the public arena.

Self-perceptions

We begin with the scientists' own assessment of their success in the political realm.
The scientists who participated in the ABM controversy felt they were efficacious
in the resolution of the issue. Seventy-two percent of the respondents were of the
opinion that scientists had been very influential in the ABM issue; 10% thought
scientists had played no influential role; and the remainder ascribed "some" or
"little" influence to the scientist participants.

Scientists, to begin with, could take credit (or blame) for the very invention of
an antiballistic missile system. Furthermore, they were judged by their peers with
having devised the concept of a light area defense.

Since the army and the Joint Chiefs of Staff were openly pushing for a heavy or
thick urban system in 1966 or 1967, when the deployment of a light area defense was
announced in 1967, anti-ABM scientists were thought by their colleagues to have been
instrumental in stopping the more extensive deployment.

The ABM participants themselves assumed that scientists had been consulted on
the Sentinel decision. Only 13% of the ABM sample thought that the decision was made
without any input by the scientific advisory apparatus. Similarly, with the Safeguard
decision, only 15% of the sample thought that no scientists had been consulted. A
wide chasm lies between being consulted and actually influencing decisions, however.
To offer a few stepping-stones across this gap, let us turn now to the perceptions of
the decision makers.

Perceptions of Decision Makers

In addition to interviewing 152 scientists who were ABM participants, the investigator talked with 70 key nonscientists involved in the debate. These included senators,
congressmen and their aides, present and former officials of the State Department,
the Arms Control and Disarmament Agency, the Defense Department, the National
Security Council, and the Armed Services, as well as ABM contractor personnel.
Analyses of the data compiled from these interviews show striking contrasts between
the perceptions of these decision makers and the perceptions of the scientists concerning the influence of the scientific community on ABM decisions.

To begin with, none of these 70 nonscientists attributed any influence to scientists
or to scientific advisory panels in the making of the Sentinel decision. The next decisions concerned the location of the missile sites near large metropolitan areas. The
data indicate a lack of influence or power attributable to any scientists other than those
officially charged with carrying out such policy decisions, that is, John Foster and
Daniel Fink in DDR&E.

On the Safeguard decision, it appears that "no-ABM" views were being expressed inside as well as outside of the new administration, but they were not considered as viable options by the men involved in preparing the Nixon adminstration's ABM proposals.

Apparently, Henry Kissinger asked some of his scientific friends in Cambridge to provide him with several papers on the technical feasibility of antiballistic missile defenses and to give him a reading of the sentiment among the scientific community in Cambridge. Late in February 1969, Kissinger was informed that a hard-point defense would win at least a 50-50 acquiescence of the scientific community. One interpretation of this episode is that Kissinger may have felt he had a signal to go from an area to a Minuteman defense. Another interpretation could be that Kissinger's sounding of the scientific community was a goodwill gesture, not intended to be taken seriously if it ran counter to the decision as it was evolving, and useful to have if it concurred with the president's decision.

If the views of scientists outside of government were only marginally influential, what role was played by the science advisory mechanisms within the government? The PSAC Military Strategic Panel was apparently specifically asked what they thought of hard-point defense and were told the question came from the president. They wrote a document that was said to be similar to the points raised by Hans Bethe in a letter he wrote to Senator John Cooper on March 21, 1969.

Bethe (and other members of the panel) thought while "ABM defense of Minuteman sites is technically feasible and, in principle, sensible. . . . The deployment of ABM around Minuteman seems to be premature."[106] Bethe and other panel members felt there would be better means of defending Minuteman than using Sentinel parts.

PSAC, apart from this panel, was (according to a majority of the members at the time) not directly involved at all. They were apparently given a briefing on ABM at their monthly meeting in February 1969 but were not given any opportunity to raise technical questions. The PSAC Military Strategic Panel was asked for its advice, but when that advice was at variance with the proposed deployment it was disregarded. It appears that, if scientists were influential in the Safeguard decision, they were not the scientists within the official science advisory bodies.

One scientist, however, was credited by many observers as having been very influential in changing the Sentinel to the Safeguard ABM system. This was Harold Agnew who was Weapons Division Leader at Los Alamos Laboratory at the time. In a paper dated January 2, 1969, titled "What's Wrong with Sentinel," Agnew wrote that "the Sentinel system could add significantly to the survivability of our strategic missile forces, the command centers which control their use and the command structure."[107]

After listing several ways in which "technically hardpoint terminal defense is a much easier problem than that associated with area defense," Agnew pointed out that "defense installations are primarily located in areas of existing military bases, thus minimizing problems presently being posed by citizens worried over safety matters or angered by dislocation problems."[108]

The Public Arena

In the public arena it was, as we have seen, primarily the Outer scientists who were the chief actors. Their influence varied from Chicago, where all the available evidence points to the Argonne scientists as the prime movers in arousing public interest in the ABM issue, to other locations where they seem to have played a lesser role.

In Seattle, the mayor had been hoping for many years to obtain most of Fort Lawton for a city park when that fort was to be declared surplus property on July 1, 1967. When the army obtained a halt to the surplus action in anticipation of using the fort for a Sentinel missile site, the mayor at first was a lone figure in his efforts to counter this action.

Soon, however, a far-flung coalition was formed, not to oppose the deployment of an antiballistic missile system per se but to oppose the proposed location of such a site at Fort Lawton. The University of Washington scientists were one component of this coalition, and their main contribution was to provide detailed data that "the Sentinel strategically can go at sites other than Fort Lawton."[109]

In Boston, each of five proposed northern suburban locations for the missile site also aroused citizen protest. Initially, in 1968, the scientists who were to be so vocal half a year later were not heard from. But, by the time of the Reading, Massachusetts, meeting on January 29, 1969, the local scientists leapt to the forefront of the battle and remained there until the Senate vote in August.

In all other locations where opposition to ABM was voiced, such as in Detroit, San Francisco, Pittsburgh, Montana, and North Dakota, Outer scientists were active in the opposition. Their role appears as a necessary but not sufficient ingredient in turning the ABM into a national issue. That is to say, Outer scientists often were the catalysts, but it was the massive citizen opposition that carried the political weight.

The Congressional Picture

In analyzing what effect the scientists' efforts had in the Halls of Congress, we are really asking how much influence can be attributed to the ABM scientist-participants in the 50-50 tie vote in the Senate in August 1969. Although Congress consists of two houses, almost all the scientists' energy was expended upon influencing the decision in the Senate.

In 1968, the first year that Congress voted to authorize and appropriate funds for the deployment of Sentinel, scientists played only a peripheral role in the five-day Senate debate that ended with the defeat of the first Cooper-Hart amendment by a vote of 34-52.

Shortly after the 91st Congress convened in Washington in 1969, the first of the anti-ABM scientists testified before a congressional committee. It was only during this brief period, in the spring and summer of 1969, that Inner and Outer scientists worked for and against deployment simultaneously. Inner scientists were collectively spending hundreds of hours on Capitol Hill, both testifying before congressional committees and explaining to senators the intricacies of phased array radars, blackout effects of nuclear explosions, and the blast resistance of Minuteman silos.

At the same time, Outer scientists were busy lobbying on all levels—local, state, and federal. These two concurrent enterprises succeeded in making ABM front-page news throughout the country for nearly six months and culminated in the 50-50 vote in the Senate in August.

How did the senators and their aides, the targets of these intensive efforts, feel about the effectiveness of the scientists' effort? The data point to an interesting symmetry between the perceptions of pro-ABM senators and pro-ABM scientists and anti-ABM scientists and senators.

In our analysis of the scientists' attitudes toward Congress and the ability of senators to understand the technical issues involved, the data indicated that pro-ABM scientists felt that the technical issues inherent in the ABM were of such complexity that nonscientists, even senators, were incapable of comprehending them. Anti-ABM scientists were of the opinion that the technical questions were not paramount and could be understood by any intelligent laymen if a conscientious effort was made by both the explainer and the listener.

These feelings seem to be reciprocated by the senators. The pro-ABM senators did not attribute a great deal of influence to the scientists' efforts. They were inclined to agree with Jeremy Stone's observation that "where you stand depends on where you sit." That is, they saw the scientists on both sides of the ABM question as "having a line to sell," as advocates, not impartial, objective witnesses. Since for each anti-ABM spokesman they could listen to a pro-ABM scientist, they tended to equate the two sides and shrug off the scientists as "lobbyists no different from any others."

The anti-ABM senators, on the contrary, were emphatic in their assertations that "without the scientists, we never could have made a close fight of it. They were informative, influential, and effective as opinion makers."

In summary, the scientists' influence in the political arena in connection with the ABM issue was not of a consistent nature. In the early days, before the deployment decision, it was scientists who were chiefly responsible for the technological advances and improvements which made antiballistic defenses seem feasible. The decision to deploy, which was a political and not a technical one, was made against the advice of the president's official science advisory apparatus but with the approval and concurrence of the science advisory bodies within the Defense Department. Influence in the location of the missile sites can be attributed to about two or three scientists working within the Defense Department's Directorate of Research and Engineering.

Outer scientists were influential in spearheading citizen opposition to the location of missiles near large metropolitan areas. It was their success in fostering this opposition that led, in our opinion, to the Nixon administration's decision to proceed first with defending Minuteman bases. Had the scientists not been able to enlist public support, it seems doubtful that the technical misgivings of the anti-ABM scientists about the feasibility of ballistic missile defenses would have become known publicly at all, or had any influence. It was the people's voices, not the scientists', that were heard in Washington.

When the scene shifted to Congress, anti-ABM senators found anti-ABM scientists useful in providing what some have referred to as the "fig leaf" function of science advisors. They provided anti-ABM senators with the technical jargon, facts, and figures to bolster their own doubts or misgivings about the system. Because the technical and political views of the anti-ABM scientists coincided with the political views of the anti-ABM senators, a situation arose where the appearance of influence was perhaps greater than the reality.

Pro-ABM scientists provided the same "fig leaf" legitimizing function to pro-ABM senators. It is perhaps oversimplified, but not inaccurate, to say that in the ABM issue, scientists were influential when their views corresponded to those of the decision makers who were at the critical decision points in the policy-making process. When the scientists' advice ran counter to these streams, it was largely unheeded, except when the scientists were able to arouse and sustain widespread public interest and participation.

Lessons to Be Learned

If we extrapolate the findings from the ABM issue to other realms, it would appear that one lesson to be learned is that scientists will be most effective in the political arena when they take their case to the people and build bridges to close the gap between "science" and "people." If scientists wish to become politically more effective, they can no longer consider "politics" a pejorative word.

Second, scientists should begin to perceive that their political concerns should not be exclusively based on the executive branch of government. The new alliances between senators, their staffs, and scientists forged during the ABM debate need to be nurtured. To encourage and facilitate mutual understanding between scientists and legislators, we might recommend that each scientific society establish two-year postdoctoral fellowship-traineeships to enable young scientists to serve as legislative or staff aides. Such fellowships would fulfill several functions: (1) they would provide young scientists with opportunities to gain firsthand knowledge, understanding, and tolerance of the political process; (2) they would begin to provide the legislative branch with its own, independent sources of expertise on the many technological issues facing it; and (3) they would provide the peer recognition and rewards for advisory services to the legislative branch which are now awarded only for services rendered to the executive branch.

For scientists unable or unwilling to undertake a two-year commitment for full-time public service, we would suggest that Congress establish summer institutes, based on the Jason-IDA model. Such summer studies on conversion from a war to a peacetime economy, disarmament, antisubmarine warfare, or other issues on which Congress feels it needs detailed technical information could be commissioned by joint, select, or standing congressional committees. Additionally, such summer institutes would provide a means for younger Outer scientists to acquire the expertise and experience about some of the nation's urgent problems they now find so difficult to obtain.

From the ABM issue, it appears that another crucial lesson is the need to build up alternative sources of research and analysis of the sort that are now available to the Defense Department. Independent, multidisciplinary institutions, similar to the Brookings Institution, and with reports publicly available, should be seriously considered. Having several competing research institutions offers a better chance of excellence but does not guarantee it. For research institutions to serve a useful function, at least two conditions are necessary: recruitment of first-rate minds and a willingness on the part of decision makers to give serious considerations to their findings.

All these suggestions have the common goal of seeking to enhance the interactions between scientists and decision makers. We feel it is imperative for younger scientists to be given opportunities to bring fresh insights and new ideas to the many urgent problems that beset us. If the ABM issue kindled the sparks necessary for more effective interactions between scientists and politicians, this may be its most lasting contribution.

NOTES

1. Authorization for Military Procurement, Research and Development, Fiscal Year 1971, and Reserve Strength, Committee on Armed Services, U.S. Senate, 91st Congress, 2nd Session, May 19, 1970, p. 2225.

2. Alice Kimball Smith, A Peril and A Hope (Chicago: University of Chicago Press, 1965), p. vi.

3. C. P. Snow, The Two Cultures: And A Second Look (New York: The New American Library, 1959).

4. Ralph Lapp, The New Priesthood (New York: Harper & Row, 1965).

5. Don K. Price, The Scientific Estate (Cambridge: Harvard University Press, 1965).

6. Harvey Brooks, "The Scientific Adviser," in Scientists and National Policy Making, Robert Gilpin and Christopher Wright, eds. (New York: Columbia University Press, 1964).

7. Spencer Klaw, The New Brahmins: Scientific Life in America (New York: William Morrow, 1968).

8. Robert C. Wood, "Scientists and Politics: The Rise of an Apolitical Elite," in Scientists and National Policy Making, Gilpin and Wright, eds.

9. Wallace S. Sayre, "Scientists and American Science Policy," in ibid., Gilpin and Wright, eds.

10. Lester W. Milbrath, Political Participation (Chicago: Rand McNally & Co., 1965), p. 18.

11. American Science Manpower 1968, a report of the National Register of Scientific and Technical Personnel (Washington, D.C.: National Science Foundation, 1969)..

12. Gabriel A. Almond and Sidney Verba, The Civic Culture (Princeton: Princeton University Press, 1963); Bernard R. Berelson, Paul F. Lazarsfeld, and William N. McPhee, Voting (Chicago: University of Chicago Press, 1954); Angus Campbell, Gerald Gurin, and Warren E. Miller, The Voter Decides (Evanston: Row, Peterson & Co., 1954); Robert E. Lane, Political Life (New York: The Free Press, 1959); Paul F. Lazarsfeld, Bernard Berelson, and Hazel Gaudet, The People's Choice (New York: Columbia University Press, 1944); Daniel Lerner, The Passing of Traditional Society (New York: The Free Press, 1959); Milbrath, Political Participation.

13. Milbrath, Political Participation, p. 16.

14. Ibid., p. 17.

15. Herbert L. Costner, "Criteria for Measures of Association," American Sociological Review, Vol. 30, June 1965, pp. 341-353.

16. Robert S. McNamara, "Address Before United Press International Editors and
Publishers," San Francisco, September 18, 1967, News Release No. 868-67, Office
of Assistant Secretary of Defense (Public Affairs), Washington, D.C.

17. United States Congress, Congressional Record, Vol. 114, October 2, 1968, p.
S 29186.

18. Interview with Newell Mack, April 10, 1971.

19. Newell Mack, "Missile Defense: A First Glance," unpublished paper, July 28,
1967.

20. Letter, Hans Bethe to Newell Mack, dated August 21, 1967.

21. "Scope, Magnitude, and Implications of the United States Antiballistic Missile
Program," Joint Committee on Atomic Energy, Congress of the United States, 90th
Congress, 1st Session, November 7, 1967, p. 135.

22. Letter, Newell Mack to Hans Bethe, dated December 9, 1967.

23. Letter, Senator John Tower to Newell Mack, dated March 28, 1968.

24. Letter, Colonel Raymond T. Reid to Senator Herman Talmadge, dated April 22,
1968.

25. "The Sentinel ABM System," United States Congress, Congressional Record,
Vol. 114, July 11, 1968, p. 20699.

26. Elizabeth Drew, "Reports: Washington," The Atlantic, Vol. 224, December
1969, pp. 4-18.

27. Stephen Dedijer, "Can Natural Science Rescue Social Science?," talk at MIT
Physics Colloquium, February 19, 1970.

28. Edward Randolph Jayne, "The ABM Debate: Strategic Defense and National
Security" (Cambridge: Center for International Studies, 1969) p. 29. See also "A
Discussion of Nike-Zeus Decisions," Draft of a speech by Mr. Fred A. Payne, deputy
director of Defense Research and Engineering (Strategic and Defense System),
delivered to a meeting of the Brookings Institution, Washington, D.C., October 1, 1964.

29. Jayne, "The ABM Debate," p. 31.

30. Benson D. Adams, Ballistic Missile Defense (New York: Elsevier Publishing Co.,
Inc., 1971), pp. 27-30.

31. Strategic and Foreign Policy Implications of ABM Systems, Subcommittee on International Organization and Disarmament Affairs of the Committee on Foreign Relations, U.S. Senate, 91st Congress, 1st Session, March 26, 1969, p. 308 (hereafter called Gore Hearings).

32. S. E. Asch, "Effects of Group Pressure Upon the Modification and Distortion of Judgments" in Readings in Social Psychology, Eleanor E. Maccoby, Theodore M. Newcomb, and Eugene L. Hartley, eds. (New York: Holt, Rinehart & Winston, 1958), pp. 174-182.

33. A selected bibliography on advisory bodies to be found in The Presidential Advisory System, Thomas E. Cronin and Sanford D. Greenberg, eds. (New York: Harper & Row, 1969), pp. 339-357.

34. U.S. Congress, Congressional Record, Vol. 116, August 6, 1970, p. 27726.

35. Ibid.

36. Ibid.

37. Ibid.

38. Ibid., p. 27727

39. ABM, MIRV, SALT, and the Nuclear Arms Race, Subcommittee on Arms Control, International Law and Organization, Committee on Foreign Relations, U.S. Senate, 91st Congress, 2nd Session, June 4, 1970, p. 442.

40. Ibid.

41. Ibid., p. 523.

42. Ibid., p. 522 (emphasis in original).

43. Transcript of President Nixon's Press Conference, New York Times, January 31, 1970, p. 14.

44. "Aim at Clarendon, Westchester Sites for Missile Base," Suburban Life, October 26, 1968.

45. David R. Inglis, "Conservative Judgments and Missile Madness," Bulletin of the Atomic Scientists, Vol. 24, May 1968, pp. 6-11; David R. Inglis, "Nuclear Threats, ABM Systems and Proliferation," Bulletin of the Atomic Scientists, Vol. 24, June 1968, pp. 2-4; David R. Inglis, "The Anti-Ballistic Missile: A Dangerous Folly," Saturday Review, Vol. 51, September 7, 1968, pp. 26-27 and 55-56.

46. Jerome B. Wiesner, "The Cold War Is Dead But the Arms Race Rumbles On," Bulletin of the Atomic Scientists, Vol. 23, June 1967, pp. 6-9.

47. Richard L. Garwin and Hans A. Bethe, "Anti-Ballistic Missile Systems," Scientific American, Vol. 218, March 1968, pp. 21-31.

48. James Tuohy, "A-Missile Sites in W. Suburbs?," Chicago Sun-Times, November 15, 1968, p. 1.

49. "Chronology of Chicago's 'Great Missile Controversy,'" prepared by West Suburban Concerned Scientists, Argonne, Ill., November 23, 1968, Mimeo.

50. Ron Dorfman, "Expect Missile Site Decision Early in '69," Chicago's American, November 20, 1968.

51. Ibid.

52. "Missile Site Objections Mushrooming," Suburban Life, November 30, 1968, p. 1.

53. James Cannon, "Voice Opposition to Area Missile Site," Suburban Life, December 7, 1968.

54. Ibid.

55. Calvin Trillin, "U.S. Journal: Lake County, Illinois," New Yorker, Vol. 45, February 15, 1969, pp. 100-106.

56. David Murray, "May Widen Missile Shield to Cover Soviet: U.S. Aide," Chicago Daily News, December 20, 1968.

57. Trillin, "U.S. Journal."

58. John MacLean, "2300 Attend ABM Protest," Chicago Tribune, May 15, 1969.

59. "No Defense Against Attack," Bismarck Tribune, May 1969.

60. William Beecher, "Acheson Group Seeks 'Balanced' Defense Debate," New York Times, May 27, 1969.

61. Smith, A Peril and A Hope, p. 142.

62. Ibid., p. 133.

63. Ibid., p. 136.

64. James K. Batten, "Why the Pentagon Pays Homage to John Cornelius Stennis," New York Times, Section 4, November 23, 1969, p. 44.

65. Jayne, "The ABM Debate," Chapters 10, 11; Herbert F. York, "Military Technology and National Security," in Great Issues of International Politics, Morton A. Kaplan, ed. (Chicago: Aldine Publishing Company, 1970), p. 373; Gore Hearings, p. 122, telegram from Donald Hornig.

66. U.S. Congress, Congressional Record, Vol. 113, August 21, 1967, p. 23436.

67. Jacob Javits, speech delivered at New York University, November 6, 1967, reprinted in Bulletin of the Atomic Scientists, Vol. 24, February 1968, pp. 18-19.

68. Ibid.

69. U.S. Congress, Congressional Record, Vol. 114, October 2, 1968, p. 29172.

70. Ibid., p. 29175.

71. Ibid., p. 29178.

72. "Missile Foes Ask Full Inquiry," Chicago Daily News, November 22, 1968.

73. Donald R. Matthews, U.S. Senators and Their World (New York: Random House, 1960), p. 92.

74. Ibid., p. 114.

75. Letter from Cameron B. Satterwaite to Senator Richard Russell, December 26, 1968.

76. Richard H. Stewart, "Opposition Grows in Congress to $5.5 Billion ABM," Boston Globe, February 2, 1969.

77. Ibid.

78. Gore Hearings, p. 307.

79. Ibid.

80. Ibid., p. 327.

81. Ibid., p. 328.

82. Jerome Wiesner, "Comprehensive Arms-Limitation Systems," Daedalus, Vol. 89, Fall 1960, pp. 915-950.

83. Freeman Dyson, "Defense Against Ballistic Missiles," Bulletin of the Atomic Scientists, Vol. 20, June 1964, pp. 13-18.

84. Jeremy Stone, "Containing the Arms Race," Bulletin of the Atomic Scientists, Vol. 21, September 1965, pp. 18-21.

85. Philip Geyelin, "Army Plan to 'Sell' Sentinel Revealed," Boston Globe, February 6, 1969.

86. Donald Brennan, "The Case for Missile Defense," Foreign Affairs, Vol. 47, April 1969, pp. 434-448.

87. Lester W. Milbrath, "Lobbying," International Encyclopedia of The Social Sciences, Vol. 9 (New York: The Macmillan Company, 1968), pp. 441-445.

88. Matthews, U.S. Senators and Their World, p. 177.

89. Letter, Richard Garwin to each senator, July 10, 1969.

90. Norman W. Storer, The Social System of Science (New York: Holt, Rinehart & Winston, 1966), pp. 8-9.

91. Everett Carll Ladd and Seymour Martin Lipset, "The Politics of American Political Scientists," P.S., Vol. IV, Spring 1971, p. 141.

92. David Nichols, Political Attitudes of a Scientific Elite, 1968 Ph.D. dissertation, Department of Political Science, M.I.T., Cambridge, Mass.

93. Everett Carll Ladd, "Professors and Political Petitions," Science, Vol. 163, March 28, 1969, pp. 1425-1430.

94. Henry A. Turner and Charles Spaulding, "Political Attitudes and Behavior of Selected Academically-affliated Professional Groups," Polity, Vol. 1, Spring 1969, pp. 309-336.

95. Stanley Eitzen and Gary Maranell, "The Political Party Affiliation of College Professors," Social Forces, Vol. 47, December 1968, pp. 145-153.

96. McNamara, "Address Before United Press International Editors and Publishers."

97. "Text of President Nixon's Announcement on Revised Proposals for Sentinel Antiballistic Missile Program," New York Times, March 15, 1969.

98. J. C. R. Licklider, "Underestimates and Overexpectations," in ABM: An Evaluation of the Decision to Deploy an Antiballistic Missile System, Abram Chayes and Jerome B. Wiesner, eds. (New York: New American Library, 1969), p. 129.

99. John Foster, "Safeguard—A Forum of Opinion," Modern Data, Vol. 3, January 1970, pp. 58-59.

100. Computer Professionals against ABM, "Fact Sheet," Ossining, N.Y., n.d., Mimeo, page 7.

101. "Information Bulletin" (Internal Employee Information Bulletin), Bell Telephone Laboratory, Murray Hill, N.J., May 15, 1969.

102. Robert K. Merton, Social Theory and Social Structure (New York: Free Press, 1957), p. 160.

103. Albert Wohlstetter, letter to author, May 18, 1971.

104. Wood, "Scientists and Politics."

105. Sayre, "Scientists and American Science Policy."

106. Gore Hearings, p. 369.

107. H. M. Agnew, "What's Wrong With Sentinel," unpublished paper, January 2, 1969.

108. Ibid.

109. Svein Gilje, "The Battle of Fort Lawton," Seattle Times, October 13, 1968.

Robert C. Wood

Professional meetings in the physical and medical sciences have been the setting in recent years for an unaccustomed range of political activism on national issues. From petitions and caucuses to confrontation and disruption, this activity has shaken the usual image of a scientific community wrapped in rationality and detached from the real world.

David Nichols' special contribution in the chapter that follows is to sort and analyze the scientists' "political interest groups" that are involved in public activism aimed at changing national policy. Adapting a system of classification that originated three hundred years ago with Hobbes' Leviathan, he divides scientists between radicals, moderates, and conservatives on the basis of their attitudes toward government weapons development and political activism among scientists. He then separates the moderate from the radical groups according to their policy outlook and political strategy. The radical groups, for example, are those whose politics includes "a criticism of the existing economic system as serving the interests of the rich at the expense of the people," and whose strategy favors direct confrontation and "economic action."

The differences, as Dr. Nichols shows, are partly generational. The older, more professionally established leadership is identified with the moderate groups such as the Federation of American Scientists and the Scientists' Institute for Public Information. While the agenda of the moderates grew out of Hiroshima and the problem of nuclear arms control, that of the radicals was shaped by the civil rights movement and the Vietnam war.

Nonscientists will be surprised at the plethora and relative longevity of overt political action by a community long thought to be immune to issues of power and prestige. From the emergence of the "pragmatic" atomic scientists at the end of World War II, to the fervent activity of the morally conscious, antiwar scientific groups of the present time, Nichols outlines the full pattern of recent scientific activism. Scientists will recognize themselves and each other but will, I think, profit more by being put in context—knowing what others were up to and finding their relative effectiveness measured and evaluated.

How one assesses the political effectiveness of scientists is highly dependent on the scale employed and on what one decides to measure. Certainly the scientific community's most striking impact on national policy has derived from individuals rather than groups, from within rather than outside the government, and from the executive rather than the legislative branch.

Had Professor Nichols written ten years earlier, in post-Sputnik days, when scientists and engineers concentrated their attention on the executive branch and international and security affairs, the depiction would have been different.

In any case, one hopes the scientists will persevere. Most political issues have a technical component; and technical issues, as Dan Greenberg has pointed out, are highly political. The era of antiseptic detachment has passed, and we cannot really mourn its passing.

David Nichols

INTRODUCTION

A bewildering array of associational political interest groups based in the scientific occupations exists today.[2] Through these groups scientists and doctors have challenged minor and major public and "private" institutional policies. Young members of the Medical Committee for Human Rights, rallying around the slogan "Health for People, Not for Profit," have gone to court in a fight against napalm sales by the Dow Chemical Corporation, and have confronted the American Medical Association at its meetings with fundamental criticism of the political economy of health. Members of the Society for Social Responsibility in Science, restricting themselves to more conventional methods of press releases and dissemination of literature, have condemned American atrocities in Vietnam. Members of the Scientists and Engineers for Social and Political Action presented Dr. Edward Teller with a "Dr. Strangelove" award (against his will) at a meeting of the American Association for the Advancement of Science. Physicists in the Council for a Livable World have counseled senators and senatorial staff on the dangers of the antiballistic missile system in C.L.W. seminars and at congressional hearings.

Such political groups are the most publicly visible means by which scientists participate in political affairs, yet no overview of the organized activism that they carry on is presently available to students of science and public affairs. It is the purpose of this chapter, therefore, to contribute an introductory analysis of this particular aspect of scientists' political activity.

Scientists

We take the membership of scientific occupations to include natural scientists, medical doctors, and most engineers (thus excluding social scientists, technicians, and laboratory workers). So understood, "science" means an interrelated set of occupations that all deal with understanding and manipulating the natural world (organic and physical), and doing so on the basis of some body of logico-intellectual principles or theory. The possession of knowledge that is critical in the functioning of contemporary society is another thing that scientists, so defined, have in common.

We can think of the practitioners of scientific occupations as members of a broad skill group whose politics can be studied. This approach has already been taken by Robert Wood in his provocative essay, "Scientists and Politics: The Rise of an Apolitical Elite."[3]

Many writers, impressed with the social importance of technoscientific skills, have projected a necessary increase in the political power of the skill group.[4] These projections are, however, metaphysical in character and not based on empirical research. They make straighforward descriptive analysis of scientist political behavior all the more necessary.

The Forms of Scientist Political Behavior

The political interest groups of science, the subject of this chapter, represent only one facet of scientist political behavior: <u>organized public activism, national in scope and aimed at changes in policy.</u> This activism should be understood to include the activities of students in the process of becoming members of the scientific occupations. But this activism should be distinguished from political activity in which scientists may participate but which has no overt relationship to membership in the scientific occupations, such as active membership in a city council or a P. T. A. or voting in elections.

Though our interest here is in political activity overtly related to membership in the scientific occupations, one very important aspect of this activity will be excluded: policy making by institutional authorities. Institutional policy making includes administrative decisions made by government officials with scientific backgrounds. Federal science policy has been studied at great length, although (particularly regarding the role of scientist administrators) in insufficient depth.[5] Governmental science advisory committees are appendages of the official policy-making process; their role has been examined elsewhere.[6]

To exclude institutional policy making from this study is also to exclude decision making by officials in professional societies. These societies do not engage in overt political lobbying as such and therefore fall without our definition of political interest group. Of course it would be wrong to think that professional societies do not shape the political economy of the scientific occupations through policies from whose effects the individual member cannot be entirely free. In addition, the current efforts of the leadership of these societies to contain internal political activism in effect help to support the present set of governmental policies by preventing them from being criticized by professional societies as such.

One might argue that the American Medical Association is an exception and is a political interest group playing an overt lobbying role. The AMA does lobby but is nevertheless excluded from this study. AMA lobbying is aimed at retaining the basic shape of the existing health economy, whereas we wish to analyze public activism that appears to challenge established policy. Moreover, AMA lobbying against various proposed

changes in social health policy is probably not as politically important as its institutional policy control over the medical occupation as such. In any case, the AMA has
been extensively studied.[7]

Public scientist activism carried out in an individual or sporadic way, or on a
strictly local basis, cannot be considered here. The complete process of separating
the political interest groups from all scientist political behavior is represented
schematically in Table 1. Table 2, which follows, overviews the groups to be studied
here.

Proceeding on the basis of the above concept of political interest groups, we can
divide these groups into two sets according to their differing political outlooks (attitudes,
positions). The most appropriate categories for contemporary scientific groups seem

Table 1. Separating the Political Interest Groups from Other Political Activity
in the Scientific Occupations

(Adapted from Thomas Hobbs, Leviathan (Indianapolis, Ind. : Bobbs-Merrill
edition, 1958) pp. 76-77.)

All political activity by members of the scientific occupations	Activity not overtly related to membership in occupations (for example, voting, P.T.A.)				
	All political activity overtly related to membership in the scientific occupations	All political activism reactive to policy, aimed at policy change	Sporadic or individual activity; activity local in scope		
			Organized activism, national in scope, aimed at policy change: the political interest groups	(Moderate groups)	Specific groups
				(Radical groups)	Specific groups
		Institutional policy making public and private; official advising			

Table 2. The Political Interest Groups of Science (Preliminary Overview)[a]

Name	Initials	Issue Focus	Political Classification	Date of Founding	Approximate Membership
American Association of Scientific Workers (defunct)	A.A.S.W.	Social responsibilities of scientists as workers	Early radical	1938	<1,000
Computer People for Peace	C.P.P	Economic Control of Technology	Radical	1968	>200
Council for a Livable World	C.L.W.	Nuclear arms control	Moderate	1962	>10,000[b]
Citizens League against the Sonic Boom	C.L.A.S.B.	Supersonic transport (SST)	Moderate	1965	5,000
Federation of American Scientists	F.A.S.	Nuclear arms control	Moderate	1946	2,000
Medical Committee for Human Rights	M.C.H.R.	Economic control of medicine	Radical	1964	<10,000
Physicians for Social Responsibility	P.S.R.	Ethical responsibility of doctors as individuals	Moderate	1961	900
Scientists and Engineers for Social and Political Action	S.E.S.P.A.	Economic control of science	Radical	1969	2,500-5,000
Scientists' Institute for Public Information	S.I.P.I.	Environmental information	Moderate	1963	20[c]
Society for Social Responsibility in Science	S.S.R.S.	Ethical responsibilities of scientists as individuals	Moderate	1949	<1,000

[a]Other groups mentioned but not analyzed here are Dentists for Peace (moderate), Health Policy Advisory Center (Health-PAC, a radical nonmembership group), Physicians Forum (radical), and Student Health Organization (radical)

[b]C.L.W. is a nonmembership group. Listed are financial supporters for a two-year period.

[c]We have no data on associate memberships as they have existed only since 1970. There are 20 full members of S.I.P.I.

to us to be <u>moderate</u> and <u>radical</u>. Actually this distinction does not come to us from
"out of the blue" but represents an adaptation to groups of distinctions found useful in
an earlier study of the attitudes of individual scientists and engineers—a study that, in
a sense, is the intellectual parent of the present investigation. It may be helpful to
devote a few words to a summary of this earlier study of individuals and then to indicate
the rationale for modifying its categories and using them to speak of "moderate" and
"radical" scientist political interest groups.

The MIT Study

In 1965, depth interviews were conducted with some three dozen of the politically most
active members of the scientific and engineering faculty of the Massachusetts Institute
of Technology.[8] The results of this study were reported in the author's MIT doctoral
dissertation.[9] The study found the terms conservative, moderate, and radical useful
in classifying political outlooks among scientists.

The definition of political activity used for the MIT study was broader than that
employed here. It included not only political interest group activity as delineated here
but also other forms of political behavior, especially government service and administration of university laboratories functionally engaged in the execution of public
policy (that is, financed by government and doing research on weapons systems for
government).

Owing largely to the broad definition of political activity in the MIT study, 24% of the
individuals interviewed were classified as <u>conservative</u> in political attitude. A conservative attitude entailed basic satisfaction with the existing long-range government policy
of substantial weapons development, combined with a general opposition to political
interest group activity in the scientific occupations. All of these conservatives had been
chosen for interviews because of their substantial experience in government service
or laboratory administration. Only four government servants included in the survey
were not conservative. Table 3 shows this and also shows that none of the scientists
included in the MIT study, because of their substantial activity in political interests
groups (political activism), were conservative in attitude.

Forty-one percent of the sample of MIT scientists were classified as <u>moderate</u> in
political attitude. A moderate attitude entailed basic dissatisfaction with the existing
pattern of armament and of military exploitation of science, tinged with optimism that
presidents were now seriously interested in arms control. This policy outlook was
combined with the view that political interest group activity by scientists had been useful
in the past in prodding administrations to consider seriously scientists' opinions. The

Table 3. Scientists Interviewed in the MIT Study, by Political Attitude and
Political Activity (1965)

| Political Attitude | Political Activity | |
	Government Service	Political Activism
Conservative	9	0
Moderate	4	11
Radical	0	13
Total	13	24

moderate attitude entailed a preference for conventional methods of expressing political views at the time of the interviews: a skeptical attitude toward vigorously oppositional action or confrontatory action.

Those classed as radicals had a different attitude toward action from that of the moderates: they felt at the time that vigorous political interest group activity was vital. Indeed, they were all engaged in organized oppositional political action perceived as placing pressure on government to change policy; several of them were involved in ad hoc local efforts that had sprung up in response to Vietnam war escalation. Thirty-five percent of the interviewed scientists held this attitude. The radical attitude entailed the same policy dissatisfaction as the moderate attitude, but with more pessimism about any commitment of the executive branch to reverse the arms race and its concomitant military exploitation of techno-scientific skills.

In addition to determining that each of the politically active scientists could indeed be classified as conservative, moderate, or radical in political attitude toward basic science-related issues, the MIT study probed opinions on many particular topical issues. These results are not especially germane to our purpose here, nor could we begin to summarize them in the space available. But there is one additional set of findings of that study which may be of interest here; this concerned the particular scientific disciplines of individuals with different political attitudes. These findings are summarized in Table 4.

The mathematicians and biologists were largely radical, while the physicists and theoretical computer scientists were by and large of moderate attitude. It should be emphasized that these results were only for one institution, and also that the study was restricted to the politically most active scientists. We do not know what the political attitudes of the other scientists were, except by inference from their relative inactivity. Nevertheless, these findings corroborate suggestions that have been put

Table 4. Politically Active Scientists at MIT, by Discipline and Political Attitude (1965)

Discipline[a]	Political Attitude of Interviewed Scientists			Total Interviewed in Discipline	Total Active but not Interviewed	Total in Relevant Faculty[b]
	Conservative	Moderate	Radical			
Laboratories Involved in Weapons "R&D"[c]	4	0	0	4	1	d
Physics Department	2	7	2	11	1	89
Electrical Engineering (Computer Science) Department	1	5	2	8	2	131
Biology Department	0	1	4	5	1	22
Mathematics Department	0	0	5	5	0	62
All Other Engineering Science Departments	2	2	0	4	3	335
Totals	9	15	13	37	8	639

[a]Research center or departmental affiliations were used as indicative of "discipline" here. In departments not specifically listed, few active scientists could be found ("Other Departments").

[b]"Faculty" were defined as full-time instructors (or above) so long as either Ph.D. degree or tenure was held.

[c]Though laboratory administrators held appointments in other engineering departments, their major professional concern was the activity of their R&D centers; thus this affiliation is used in this table.

[d]Research cadres at the laboratories were excluded from our definition of faculty. The accounting irregularity here is due to our desire to list laboratory administrators as a separate "discipline."

forth from time to time to the effect that scientific discipline may have an observable connection with political outlook.[10] Therefore, attention will be paid to the disciplinary backgrounds of political leaders in the analysis of the interest groups of science in the later sections of this chapter.

Applying Attitudinal Categories to Groups

Political groups can be seen as systems or units with histories and careers of their own.[11] It appears reasonable, therefore, to modify terms classifying individual political attitudes and adapt them to the description of group political outlooks. It is true the individuals in groups ordinarily have differing opinions about what the group should do politically, but nevertheless there is, over time, a more or less determinate group "politics" which can be discovered through research and analysis.

We shall not apply the conservative attitude classification to groups in the following analysis. The only "scientist" political group national in scope to which it would apply is the A. M. A., which, as explained earlier, is excluded from our analysis. No doubt there are substantial numbers of scientists and engineers basically satisfied with the present decision-making process with respect to the application of technoscientific skills, as well as with the maintenance of a continuing high level of military preparedness; but they have simply not formed political action groups in opposition to existing patterns of policy. They do not have to.

Moderate and radical classifications can be used. In the MIT study, what moderates and radicals held in common was a dissatisfaction with public policy, and what distinguished them from each other were their attitudes toward political action. These differences need to be refined in order to be usefully applied to scientist political interest groups.

1. There are now more clearly differing policy outlooks on the part of moderates and radicals because of the development of protest during the past eight years. What hardly existed in 1965, when protest was directed at the facts of militarization and warfare themselves, was an economic criticism of the equity of the political economy of science and its uses. But socialist or at any rate anticapitalist analyses are now an important part of student and professional protest generally.

When the politics of a group includes a criticism of the existing economic system as serving the interests of the rich at the expense of the people, its policy outlook may be termed radical. When the politics of a group includes no such critique, or at most includes the liberal call for new domestic policy priorities, its policy outlook may be termed moderate.

Table 5. Delineation of Political Interest Groups

Defining Dimension	Moderate Group	Radical Group
Policy Outlook	Political critique of government policy with respect to the application of science and technology	Includes critique of the economic system's uses of technoscientific skills; politico-economic policy critique
Political Strategy	Traditional lobbying methods; political actions	Includes direct confrontational action; economic and political actions

2. The conceptions of moderate and radical attitudes toward political action developed in the MIT study, can be applied with less modification to scientist political groups. Groups restricting themselves to conventional forms of political action—press releases, literature dissemination, and other public education efforts; congressional testimony, campaign contributions, or similar lobbyist efforts—are considered moderate in terms of their strategy for political action. Groups that support sit-ins, confrontations at meetings, or any of a variety of other forms of direct action aimed at officials of laboratories, hospitals, universities, or professional societies that are held to be tied in the larger political system—these are radical in terms of their strategy for political action even if they also issue press releases and print literature as do the moderate groups.

Table 5 restates the distinctions between moderate and radical scientist political interest groups. The two sets of groups can be analyzed separately, beginning with the moderate groups.

THE MODERATE STYLE [12]

The Background

Even before the atomic bomb was dropped on Hiroshima in 1945, nuclear physicists were engaged in political action critical of government policy and aimed at controlling the destructive force that they themselves had helped unleash. Nuclear physics was to remain at the center of a definite political movement of scientists from then on, in the twin senses that the movement (1) focused on the dangers of the exploitation of nuclear energy in an arms race, and (2) was spearheaded by physicists.

That small early protest, involving petitions and other representations to higher
authorities from scientists working within the secret atomic bomb project (most notably
Leo Szilard), of course did not succeed in preventing military use of the atomic bomb.
But it did set the stage for an impressive postwar public campaign around the issue of
governmental organization of nuclear energy. [13] When an administration-backed bill
for the domestic control of atomic energy was introduced in Congress in the fall of 1945,
a number of local political associations of scientists and engineers had already sprung
up in anticipation of critical decisions to be made regarding domestic and international
control of nuclear energy. These groups were located principally at atomic bomb project
sites, and their formation had been spurred by the dramatic impact the bombings of
Hiroshima and Nagasaki had on many young scientists.

The War Department bill of 1945 was a shock to the politically concerned younger
scientists, for it contradicted assurances from J. Robert Oppenheimer as well as
other scientists serving government as administrators that the draft legislation would
be agreeable. The bill proposed that atomic energy be controlled by a part-time commission—insulated from executive authority, including four military service representatives among the nine members, and appointing its own administrator. To this
the scientists objected on the grounds that it could produce military control of the
resource. The bill provided rigid control, virtually a monopoly, on nuclear research,
and specified drastic security penalties for the release of information. To this the
scientists also objected. To fight the bill, the scientists organized their several groups
into a Federation of Atomic Scientists.

Soon a broader Federation of American Scientists (F.A.S.) was established, and it
eventually supplanted the first federation. In their fight against the War Department
bill the F.A.S. activists successfully obtained press coverage of their resolutions,
petitions, testimony, press releases, and press conferences. Sympathetic radio commentators and columnists publicized the scientists' desire for international control
of nuclear energy as well as their opposition to the bill. An article by a group of physicists appeared in Life.

Several nonscientist groups had expressed interest when the scientists first opposed
the atomic energy bill. To encompass them, the federation established a separate
National Committee on Atomic Information. The scientists also had continuing communication with sympathetic politicians, several of whom where independently interested
in the bill. There were sources of opposition to the bill in Congress and in the executive branch, without which the scientists would have been ineffectual.

The coalition of which the "scientists' lobby" was a part achieved some success. The administration supported a different bill. In the end some changes in the atomic energy legislation were effected (though the final bill was not as different from the first bill introduced as the scientists had hoped). The Atomic Energy Commission was finally constituted as a full-time, wholly civilian, and presidentially responsible body, with a general manager also appointed by the president. In addition, security provisions were slightly less repressive than in the first bill. However, with the failure of the government to pursue international control negotiations seriously with the full development of the cold war, nuclear energy was destined to be used mainly for weapons development despite the modest victory of the scientists' lobby with respect to the structural organization of the A.E.C.

The policy outlook of the scientists fighting for peaceful application of atomic energy in the 1945-1946 struggle had been foreshadowed in the Franck Report of early 1945. The Franck Report was not an official document but an amazingly prescient report drawn up in Chicago by a group of scientists (chaired by James Franck) at the Metallurgical Laboratory of the atomic bomb project. The report argued that the only alternative to international control of nuclear energy was an arms race that ultimately could afford no military security to the United States. Fearing that use of the bomb against Japan would destroy the mutual trust among nations that they held to be the basis of effective international collaboration, the authors advocated instead " a demonstration of the atomic weapon...before the eyes of representatives of all the United Nations, on the desert or on a barren island."[14] Dropping the bomb on Japan would be "military expediency," whereas the concern of the scientists was "long-range national policy."[15]

The Franck Report's themes—the desirability and necessity of effective international control of nuclear weapons, and the danger of treating scientific resources as a national military tool—were central to the politics of the scientists' movement from the very outset and remain so today. A number of associated themes—civilian domestic control of atomic energy and scientific resources, freedom in scientific research, and government support for basic research—remained a part of the arms control-centered movement from the postwar period to the present.

The early F.A.S. did not have a highly developed political analysis of different group, class, or ideological interests determining governmental decisions about nuclear energy in the United States, the U.S.S.R., or any other country. Instead, a fundamental policy implication was drawn from the fact of technology itself. If governments tried to develop atomic energy for military ends, it would be self-defeating owing to the nature of the technology. The only political assumption that had to be made was that the U.S.S.R. or perhaps other states would try to offset U.S. military stockpiling. An explicit normative assumption was that scientists, knowing the technology best, had a special responsibility to engage in political action.[16]

The political action in which scientists felt they ought to engage essentially involved educational efforts—efforts to educate citizens and politicians on issues with a scientific component. The strategy presupposed a democratic political process and, beyond that, presupposed that decision makers could be moved by correct information and/or public opinion informed by correct information.

Both the strategic and policy outlook of the F.A.S. remained the same over the years. This was true even after the F.A.S. in the late 1940s accepted cold war military mobilization as valid. Most other scientist political activism (at least until the mid 1960s) resembled the politics of the F.A.S. F.A.S. activism died down after 1946, and there never recurred quite so vigorous a scientist political effort until the 1969-1970 ABM debates.

There is a history of scientist activism between the "scientists' lobby" and the ABM debates, although it cannot be detailed here. The moderate issue-oriented cast given scientist politics by the postwar legislative battle continued to characterize practically all of this activism. In 1949, the Society for Social Responsibility in Science was formed; it consists of scientists and technicians willing to draw some line between work they will and will not do based on ethical judgments about what is good, and what is destructive, for people. In the 1950s a new critique of the arms race centered on the dangers of radioactive fallout from nuclear weapons tests. Many scientists were active through the National Committee for a Sane Nuclear Policy, which was not, however, a scientist group. In the McCarthy era, fights to protect the reputations of victimized scientists were mounted, although the security system as such was not fought—only its abuses, which were plentiful. In 1962 the Council for a Livable World was established as a scientist-led lobby of those who would contribute money to candidates committed to peace. Physicists were always prominent in the leadership of political activities dealing with arms control, and these activities were central to scientists activism in general during the period.[17]

The Federation of American Scientists[18]

The F.A.S. is today still the closest approximation to a general national political lobby based in the scientific occupations. Like many issue-oriented groups, the main emphasis of F.A.S. activities is on influencing congressmen directly, by maintaining liaison with members of Congress or their staffs, and indirectly, by shaping the public opinion environment in which legislators operate.

The F.A.S. had become a weak and not very active association when, in 1970, it was chosen by prominent scientists who had become active in the arms control movement as an organizational vehicle for intensifying the struggle against armament policies of the Nixon administration—centrally the ABM policy. Herbert York, who after serving as director of Defense Research and Engineering in the Eisenhower administration, had become more critical of U.S. weapons policy, became chairman of the group.[19] Jeremy Stone, a professional arms control expert, became full-time director of the F.A.S.[20] Some of the new leaders had been members of the F.A.S. and some had not, but their assumption of key positions was welcomed by the leaders at the time; that is, it was not the result of any political struggle in the group. With the reorganization, the F.A.S. had a full-time director and functioning national office for the first time in many years.

The political energy of the F.A.S. remains focused on the problem of nuclear arms control. The group opposed deployment of the ABM system. It urged an agreement at the Strategic Arms Limitation Talks to forestall the MIRV (multiple independently targeted reentry vehicle) modification of existing Soviet and American intercontinental ballistics missile systems.

Most of the organizational efforts of the F.A.S. continue, at this writing, to go into press releases and congressional testimony. Scientifically informed criticism of the rationale of strategic weapons systems and information dealing with problems of inspection in any arms control agreement constitute a cornerstone of F.A.S. public relations strategy. The prominence of the scientists in the F.A.S. leadership—or those willing to associate themselves with F.A.S. statements—was such that during the ABM debates the New York Times, the Washington Post, and other major papers did print stories utilizing the releases. The F.A.S. maintained some legislative contacts during the ABM debate, principally through the staffs of potentially anti-ABM senators. This access was used to deliver summaries of congressional testimony on the ABM and MIRV to senatorial offices.

The anti-ABM fight itself was organizationally managed through ad hoc formations, particularly the National Citizens Committee Concerned about Deployment of the ABM (an "establishment" group headed by former Deputy Defense Secretary Roswell Gilpatrick and former Supreme Court Justice Arthur Goldberg) and the counterpart National Science Advisory Committee on the ABM, headed by former presidential science advisor

Donald Hornig and Herbert York. [21] Supplementing York's activities, the F.A.S.'s
director and another of its leaders, Barry Casper, spent several summer weeks in
full-time anti-ABM lobbying. Thus the anti-ABM lobby and its senatorial members
benefited from the F.A.S., which played a helpful "backup" role. (Anne Cahn's essay
in this volume discusses the collaboration between senators and a number of very
prominent scientists in the ABM debates.)

F.A.S. is in a state of transition. Its leader have ambitious plans: to multiply the
membership; to establish chapters at universities around the country; to establish a
network of scientists and engineers who, in each congressional district, will contact
the local representative with an eye to influencing him through sound information about
arms control and other matters. [22]

Despite the visibility of the F.A.S. during the ABM fight and despite its continued
activity, we have found no evidence of a sustained higher level of effectiveness by the
group in 1971. F.A.S. remains only an approximation to a general political lobby
based in the scientific occupations—and, in fact, not a very close approximation. In
March 1971, its membership still hovered about the two thousand mark, as it had for
several years, despite many mailings to lists of potentially interested persons. Since
it has no mass membership, it cannot persuade congressmen with the power of the votes
of its members. Its budget is markedly smaller than even that of a similar organization,
the Council for a Livable World.

There are several reasons why the appeal of the F.A.S. to a broader scientist constituency remains limited. First, the group has not broken out of its one-issue mold.
It is true that positions are taken on diverse problems concerning science and society,
such as environmental pollution and conversion of industrial plants from military production. The F.A.S. has also directed public criticism at government inroads into
scientific freedom, more frequently in the McCarthy era than today. The consistent
focus of F.A.S. public relations work, however, remains the strategic nuclear arms
problem. In a period when U.S. counterrevolutionary conventional warfare in Vietnam
has had so much to do with the increase in the political awareness among students and
scientists, the constancy of this concentration on the nuclear issue is striking. The
chairman of the group is a physicist, and over half of the F.A.S. leaders are physicists.
While physicist domination may decrease, leadership continuity will probably be sufficient to assure that the political activities of the group remain largely "locked" in to
the concern with the issue of nuclear arms control which occasioned the formation of
the F.A.S. twenty-five years ago.

Second, since the political outlook and action strategy of the group is decidedly
moderate, it cannot appeal to those younger scientists frustrated with the failure of
conventional scientist lobbying. Conventional lobbying—advice to congressmen, public
relations work—is the only present or projected form of F. A. S. action. The F. A. S.
critique of the arms race is trenchant enough, but its premise is that governments
can be shown that it is in their interest to negotiate an arms control agreement—certainly not that mass pressure is needed to force steps to demilitarization.

The policy outlook is liberal, pragmatic, balanced. The government should "balance
the risks of a continuing arms race with the often exaggerated risks of arms limitation";
the F. A. S. recognizes "the necessity for a personnel security program" but opposes
"application of irrelevant criteria and arbitrary procedures."[23] Comments by the new
executive director exemplify the policy outlook:

... [Those] concerned with improving American society have focussed with too great
emphasis upon ills that can be cured by reforming political structure.
... [Many] of the ills which we do attack might be better resolved without attempting
changes in structure or law. There is nothing wrong with the political and economic
structure of this country that could not be fixed if its institutions stopped operating
in such a mindless way. Corporations can be run for the public interest rather than
in the narrow economic interest; they would be if the individuals running them felt that
this was expected of them. Why not begin to expect it of them? Such expectations are
far more effective than law and they may be easier to generate. [24]

The assumption which some would characterize as naïve, that correct expectations
—rational as opposed to mindless—can change institutional policy runs through F. A. S.
political work. The similarly parallel view of corporate interests may be the key to the
absence of any F. A. S. political analysis, liberal or radical, of the relation of capitalism to militarism in American society (the "military-industrial complex").

The final reason why the appeal of the F. A. S. to members of the scientific occupations
remains limited probably lies in its apparent lack of interest in "bread-and-butter"
issues. The F. A. S. sponsored a symposium on the Ph. D. surplus at the A. A. A. S.
meeting in late 1970 but as yet has no program of action or set of recommendations
on the problem of technoscientific unemployment.

Despite its political weaknesses, the F. A. S. is the oldest, strongest, and most plausible candidate for the functional role of articulator of political interests among scientists. That the other groups to be discussed in this analysis are lesser than it is a commentary on the political disorganization of scientists, at least insofar as political interest groups are concerned.

Our analysis of the contemporary politics of the F. A. S. does not detract from the
normative value of the efforts of members who aim to change policy in a constructive

way. But even a preliminary scholarly overview cannot properly be an uncritical account. This point applies to groups analyzed in the remainder of this chapter as well.

The Council for a Livable World[25]

The Council for a Livable World was organized by the nuclear physicist—turned biologist—Leo Szilard when he felt the danger of nuclear war was becoming increasingly real. Since its foundation in 1962, the C. L. W. has in principle (and in practice) concentrated its political energies on the nuclear arms race. The main activity consists in the channeling of campaign contributions from supporters of the C. L. W. to senatorial candidates recommended by the group's leaders. Candidates to be supported are selected on the basis of their attitudes toward the arms race.

The C. L. W. is unwilling to think of itself as a membership organization. Its leaders accept contributions from "supporters" for its own operating costs or, usually, for channeling to the selected candidates. By C. L. W. accounting, over ten thousand persons gave some funds through the organization in 1968 and 1969. Wealthy supporters make some very big gifts, although most monies are raised through smaller contributions. Total funds raised in 1969 and 1970 combined were $300,000, down from a peak support level prior to the 1968 elections.

The C. L. W. distributes a newsletter to its supporters analyzing past and future campaigns. Estimates of the election chances of various candidates form an important part of the calculus of decisions as to whom to support. Reports about the victories and losses among council-supported candidates are used by C. L. W. leaders to suggest the importance of the contributions to the success of the recipients. In the nature of the case, however, it is difficult to trace the influence of contributions simply on the basis of election returns, which are what appear in the newsletter. In 1970 the C. L. W. channeled 85% of its contributions into seven major campaigns. Senators Hart (Democrat of Michigan), Montoya (Democrat of New Mexico), and Moss (Democrat of Utah), all won (each had received substantial C. L. W. support in 1964, too). Democrats Wendell Kay (Alaska), Frank Morrison (Nebraska), Albert Gore (Tennessee), and Philip Hoff (Vermont), all lost their election campaigns. Republicans have also been supported. Perhaps the clearest claim of a C. L. W. contribution to an important political victory was made by Vice-President Spiro Agnew:

This Council holds alarming leverage over some members of the United States Senate. Take my good friend, Senator McGovern, for example. In 1962 this lobby slipped Mr. McGovern $20,000 to spread around South Dakota in his campaign and since Mr. McGovern came sliding home with a 600-vote wafer-thin vote margin, it can be fairly said that the Council for a Livable World won George McGovern his seat in the Senate of the United States.[26]

Despite Agnew's attack, the C. L. W. as such did not become a major issue in the 1970 campaign. The C. L. W. has been attacked on other occasions from the political right. A political scientist would, however, have to remain skeptical about the effectiveness imputed to the C. L. W. by its critics on the right (in the absence of studies of the particular campaigns involved).

In the 1970 campaign, the C. L. W. directed about 15% of its contributions to seven other candidates, several in states judged too large for support of the sort the group could afford to make any political difference. The newsletter also comments on candidates with favorable attitudes toward the arms race though they may not have been selected for financial support. Occasionally, candidates for the House of Representatives have been supported. The newsletter also discusses military and foreign policy issues, and the C. L. W. prints informational literature. Public conferences on aspects of the arms race have been held. The C. L. W.'s director is registered as a lobbyist on Capitol Hill.

The C. L. W. has been able to maintain continuing contact with a small number of senators, directly or through their staff, by means of a series of private seminars. Speakers at these seminars are frequently scientists who have served government yet are somewhat critical of its arms control stance, such as George Kistiakowsky (presidential science advisor in the Eisenhower administration) and Hans Bethe. During the anti-ABM fight some of the meetings between eminent scientists opposing the ABM and senators leading the legislative struggle (like John Sherman Cooper, Republican of Kentucky, and Philip Hart) occurred within the framework of C. L. W. seminars. Since 1962, some 60 senators have participated in C. L. W. seminars at one time or another. Most of the seminars have been on the ABM.

While the F. A. S. holds elections to select its officers, the C. L. W. is controlled by a self-perpetuating council of six scientists. The half-dozen select the policy-making board of directors. Most members of the board are not scientists, although the largest single group, occupationally, consists of university scientists. The chairman, William Doering, is a Harvard chemist. Heading the list of officers (as president) is Bernard Feld of MIT, whose experience as a physicist dates back to the atomic energy project of World War II. Before its 1970 reorganization, Feld was also president of the F.A.S.[27] It seems fair to characterize the C. L. W. as a scientist political interest group, particularly in view of the makeup of the controlling council.

The C.L.W. is similar to the F.A.S. in its emphasis on influencing Congress; it differs in concentrating on direct influence of senators (through the contributions and seminars) rather than on shaping the public opinion environment in which legislators operate. The C.L.W.'s explicit rationale, set out by Szilard in 1962 and reiterated periodically by leaders since, is that cash contributions help congressmen find the courage of their convictions. This seems less naïve a conception than that which underlies F.A.S. public relations work, yet the magnitude of the effort which it implies is vast. Perhaps that is why Szilard envisioned a lobby of 25,000 to 150,000 persons contributing 2% of their income annually. At 150,000 members, he estimated annual contributions of $20 million per year.

Campaign financing is so expensive that Szilard's estimates were probably realistic goals for any lobby aiming to attain any real influence in the Senate as a whole. Whether he seriously anticipated reaching them is not clear. The C.L.W. has actually channeled a million dollars to the several recommended candidates since its foundation. Certainly there are lobbies that operate on smaller budgets. But, while the $38,000 received by the average senator supported in a "major" way in 1970 was not inconsequential, the C.L.W. faces problems many lobbies do not face—including the fact that there are competing and antagonistic lobbies supporting armament, representing the interests of corporations in the military business.

But the major problem for the C.L.W. as a lobby or political interest group is that it addresses itself to a broad and basic policy of government. A trade or business lobby can concentrate on key committee or subcommittee chairmen, thereby dealing with the most powerful legislators for a given policy area of relatively narrow scope. However the C.L.W. needs to influence not a few key legislators but really as many senators as possible. This springs from the fact that the policy initiative in foreign and military decisions does not lie with any one committee or even any set of committees in Congress. Rather it is squarely in the presidential domain. To seize the policy initiative, consequently, would require nothing less than an organized congressional revolt. Both the F.A.S. and C.L.W. aim at influencing congressional attitudes, but neither group has set out a political analysis demonstrating the political efficacy of combating the arms race in Congress.

The initiative in foreign and military policy formulation has remained within the executive branch of government, irrespective of administration, for the entire length of the cold war. It may be that C.L.W. and F.A.S. leaders hope that presidents wish to end the arms race and merely await the creation of a favorable congressional atmosphere before doing so. If this is the hope, it surely needs to be made explicit and defended intellectually, for its correspondence to reality is far from self-evident. Both the definition of the security "threats" necessitating military preparedness and the process of seeking out a limited accommodation with the Soviet Union and China in certain areas have been the result of closed debate in the advisory circles of the presidency.[28]

Congressional criticism of the conduct of the arms race by the president has not had much of an effect thus far. No imminent change in the policy docility of Congress can be deduced from the ABM struggle. For one thing, the ABM opponents did lose. But even a senatorial victory for the opponents of the ABM would not have affected the pattern of the arms race in any substantial way unless it then led, somehow, to a determination if the president to seek international agreements with other heads of state.

C. L. W. political activity is, then, based on the assumption that Congress, particularly the Senate, does or could shape the basic foreign-military policy outlook of incumbent administrations. A corollary assumption is that the governments (incumbent administrations) of nuclear states have no conflicts of interest that preclude their collaborating to decelerate the arms race. This second assumption is shared by the F.A.S. and by many arms control theorists, yet it implies that somehow the whole cold war was an error governments could have avoided.[29] The difficulties with this assumption can be seen in the C. L. W.'s handling of the issue of the Vietnam war.

The C. L. W. became especially interested in the Vietnam issue in 1967 and 1968, exploiting "inside track" contacts with senators and State Department officials to press for a peace plan drawn up by a small group of the Saigon elite who were discontented with U. S. and Thieu-Ky governmental policy in South Vietnam. The plan the C. L. W. endorsed envisioned a political settlement that would have denied the communists victory in the south. Thus (while critical of bombing attacks against North Vietnam) the C. L. W. accepted the basic goal of preventing success in the south by procommunist Vietnamese.[30]

The C. L. W.'s position on the Vietnam war was a criticism of American military action on the dual grounds that it could not succeed and that a political solution could guarantee South Vietnamese "neutrality." Under the endorsed proposal "the basic objectives of the United States would be realized."[31] The United States was criticized not for attempting to crush the guerrilla war by force but only for failing to seize the fruits of its military action wisely enough. The C. L. W. position was an implicit but strong defense of American anticommunism, an agreement with the very doctrine that led to the war.

Throughout its involvement, the C. L. W. generally saw the Vietnam war as a compartment of activity separate from the issues of the ABM and the arms race. Yet, a good case can be made for the notion that the Vietnam war in fact was no mistake but the result of a consistent commitment of successive administrations to establishing a noncommunist "democracy" in South Vietnam.[32] And this policy is hardly unrelated to the motivation for maintaining strategic arms. Nuclear weapons are bargaining tools that obviously influence Soviet and Chinese willingness to participate in such limited war situations as Vietnam, and there is no doubt that this factor has influenced the unvarying commitment of our national government over the post-World War II period to maintenance of a massive strategic force.

The C. L. W., then, does not view the problem of arms control in the full context of cold war politics and is not likely to do so in the future. The distinctive political strategy of the C. L. W. ironically depends upon the divorce of arms problems from the politics with which these problems are interrelated (in complex ways not fully sketched here). Only through such a separation can the hopeful concentration on senatorial conscience that characterizes the C. L. W. (and the F. A. S. as well) can be maintained. The self-perpetuating character of the leadership guarantees the continuity of the C. L. W.'s unique type of moderate politics.

Some Characteristics of Group Leadership

The F. A. S. and the C. L. W. are both dynamically political, and at the same time they are both concerned with a basic policy of the modern state—armament. We therefore analyze them in more detail than other moderate groups to be discussed later. To supplement our discussion of the C. L. W. and F. A. S., some attributes of the leadership of these key groups may be noted. Table 6 summarizes some relevant data.

From the first rows of Table 6, it can be seen that the leaders of these groups are professionally established, as has already been noted. Not surprisingly these men are on the whole middle-aged; the mean age of F. A. S. leaders in 1971 was 47 years, and of C. L. W. leaders, 52 years.[33] The type of doctoral degree most frequently possessed is the Ph. D.

Three F. A. S. leaders are also C. L. W. leaders. This is more leadership overlap than exists between any other groups analyzed in this chapter. Both groups now draw their leaders principally from Megalopolis, the densely populated region stretching from Washington, D. C., to Boston. In fact, 42% of them, presently live in one state —Massachusetts. And 35% of the combined leadership is at MIT or Harvard (in approximately equal numbers at each university).

Table 6. Some Characteristics of F.A.S. and C.L.W. Leadership[34]

Characteristic	F.A.S. (n = 29)	C.L.W. (n = 17)	Combined (n = 43)
Percent with Doctorates	97	65	84
Percent Listed in American Men of Science	90	82	86
Percent in Natural Sciences	76	47	70
Percent in Field of Largest Number of Leaders	59 (physics)	24 (chemistry)	42 (physics)
Percent Located in Megalopolis	76	88	86
Percent at Harvard and MIT	34	47	35

The table illustrates the not surprising dominance of natural scientists in the leadership of these groups. Although Table 6 does not so indicate, every one of these scientists is a university scientist. Four of the leaders (or 9% of the combined leadership) were trained in mathematics or the physical sciences but are now social scientists (they are not listed as scientists in Table 6). As much as 23% of the combined leadership could be considered political or social scientists, provided a broad definition is used encompassing law, magazine editorship, and political activity as well as psychology, economics, and political science. (Only one person in the entire forty-three is in humanities.)

If the forty-three leaders are taken as one group, the single most frequent scientific field is physics; 42% are physicists. Physicists dominate the leadership only in the F.A.S., however, as Table 6 shows. The second most frequent specific field is biology (16%); the third, chemistry (12%). More generally, most of the leaders are in the physical sciences; the next largest number, in the life sciences; and the third largest number, in the social sciences. This ranking works for either group separately or both together.

Other Groups

Our analysis of the moderate style has focused in some detail on only two of several moderate scientist political interest groups. The groups discussed earlier are both vigorously political and concerned with a major policy of the U.S. government:

armament, with its associated military exploitation of technoscientific skills. Because
these groups are both avowedly political and concerned with a basic policy, we place
them at the center of the moderate politics of science. Yet there are other groups
that are also nationally operative, and we can review the activities of the most important of these here.[35] Dissatisfaction with public policy (falling short of an economic critique) characterizes these groups, as does a moderate strategy for action
(that is, one emphasizing public relations, public education, and/or political lobbying).

A problem with placing one of these groups, the Scientists' Institute for Public
Information, into a category as a political interest group is that the organization
partially denies its political role. Its purpose is to make scientific and technical
information available to the public, and to that end "presentations of fact" are kept
"free from moral and political judgements."[36]

Nevertheless, the motivating conception of S.I.P.I.—that scientists have a special
responsibility to make factual information available to the public on issues with a
technical component—is clearly a political idea; and the expectation is that correct
information will lead to policy changes. Thus, biologist and S.I.P.I. Chairman Barry
Commoner hopes explicitly for "far-reaching social and political actions" to solve
deeply threatening problems of environmental pollution.[37] The "special responsibility"
conception underlying the S.I.P.I. organization relates it philosophically to the postwar scientists' movement. It is related to the arms-control-centered movement in
another indirect way—by its own history. S.I.P.I. is an outgrowth of local nuclear
information committees that sprang up in the late 1950s to publicize the dangers of
nuclear fallout. Scientists from the local committees organized S.I.P.I. in 1963.
When the fallout issue declined in importance after the partial nuclear test ban treaty,
the number of local committees declined from twenty-three to thirteen. The number
is now slowly increasing, and, since S.I.P.I. turned its attention to broader environmental issues related to local, state, and federal policy, an increasing amount of
local scientist action on environmental issues occurs within the framework of the
S.I.P.I. committees.

The activities of these local S.I.P.I. committees, composed mostly of natural scientists, are truly far-ranging. Local committees publicize dangers of nuclear radiation,
lead poisoning, air pollution, and other hazards. The Committee for Environmental
Information in St. Louis is the best-known local committee, not so much because it
is very active (which it is) as because Environment magazine is edited there. Originally Nuclear Information, the magazine now has a circulation of 25,000 and recently
became an official S.I.P.I. publication.

At the national level, S.I.P.I. is not only a clearinghouse for the local committees but also convenes its own meetings and publishes informational booklets. From the beginning, S.I.P.I. was funded by foundation and government grants; some of these monies are passed on to local committees. Environment is supported by its own government, foundation, and other group grants as well as subscriptions and contributions. Only in 1970 did S.I.P.I. begin to enroll associate members as a source of supplementary funds.

The leaders of S.I.P.I. —thirty-one directors and full members—are a professionally established group with an average age in the fifties.[38] The president is Margaret Mead. Three-quarters of the leadership consists of biological scientists, with more Ph.D.'s than M.D.'s. There is no overlap of S.I.P.I. leaders with C.L.W. or F.A.S. leaders. Headquartered in New York City, 37% of the leaders come from that state, and the single university contributing most of them is Rockefeller University (none are from Harvard or MIT, and only one is from Massachusetts). There is a broader geographic distribution than with the groups already discussed, since 17% of the leaders are in California and other locations include North Carolina, Illinois, Montana, and Ohio (among other states). In terms of general leadership, then, S.I.P.I. reflects policy concerns of university scientists in the biological sciences, a group entirely distinct from the leadership of the arms-control movement (despite S.I.P.I.'s origins).

Politically, S.I.P.I. avoids analysis of how policy can be changed. Undoubtedly public provision of information that is not made available by "responsible" government agencies is an important political act; yet the question of how the relevant policies can be altered is outside of the S.I.P.I. domain. Perhaps the grants that support the group would prove to be dependent upon continuation of this sort of political restraint. In its political reticence, S.I.P.I. resembles another group, the Society for Social Responsibility in Science.

The S.S.R.S. invites membership from scientists willing to decide on a personal moral basis what work they will not do as scientists. The emphasis is on "constructive alternatives to militarism."[39] The S.S.R.S. tries to spread among scientists awareness of the dangers and evils of war. In addition, it engages in public education work, attempting to generate media coverage of its press releases and meetings. The public relations aspect of S.S.R.S. work, both within the scientific community and in the larger community, is informed by the same doctrine that underlies S.I.P.I.'s work: the responsibility to create an informed public opinion. Because of this political education work, the S.S.R.S. can properly be considered a political interest group (rather than a collection of individual dissenters).

The recognition of the social responsibility of the scientist in his daily work could yield a radical political strategy—organized refusal to cooperate (the strike), or organized demands that institutions cease war work—but in the S.S.R.S. case it does not. Rather, some effort is made to help persons find new work when, out of individual conviction, they resign a job. The emphasis is decidedly not on political struggle. When S.S.R.S. was founded in 1949, cold war ideology was becoming more and more intense in American society, so that even the idea of a group of individuals refusing to cooperate in war preparations appeared daring. Since its foundation the S.S.R.S. has become a more respected, even sedate, organization.[40] The S.S.R.S. council did, however, condemn the war in Vietnam and call for U.S. withdrawal and, earlier (in 1969), dispatched scientists under its own auspices to study the effects of "defoliants" in Vietnam.[41]

The S.S.R.S. has not yet attained any great visibility in national public affairs. Nor has it attained any great membership, for it still numbers less than a thousand.[42] Yet it is international in scope, with members in twenty-four countries. (Most members are American, however.) At annual meetings all members may vote on policy positions and membership on the council. The S.S.R.S. differs from the other political interest groups in having foreign leaders (five of its twenty-four member council).

As might be expected (given the purpose of the S.S.R.S.), 92% of its leaders are scientists. The specific field of the greatest number is physics (29%). But fully a third of the leaders are in engineering fields, more than for any other group examined in this chapter. Over a fifth are employed in private business, also a relatively high number. Both the president, mathematician Alice Mary Hilton, and the vice-president, Earl Graham, are in business.

The S.S.R.S. leaders are a professionally accomplished group with an average age in the fifties.[43] They are concentrated in four states—New Jersey, Pennsylvania, New York, and Massachusetts—where 63% of them live.

Unlike the S.S.R.S. and S.I.P.I., the Citizens' League Against the Sonic Boom has as its raison d'etre a political lobbying effort, and unlike the lobbying efforts of the C.L.W. and F.A.S., the C.L.A.S.B.'s are directed solely at one very specific government project: the supersonic transport plane (SST). The C.L.A.S.B. began in England and Switzerland. Of the league's twenty-nine leaders (officers and national committee members), only a third are actually scientists.[44] But the league director, Harvard physicist William A. Shurcliffe, has been so important in the formation and continued functioning of the group that we consider it a scientist political interest group. Associate director is Harvard biologist John T. Edsall (also a C.L.W. leader).

Membership grew rapidly to five thousand, half of whom are students in high schools and colleges.

The campaign against the SST, in 1970 and 1971, was a politically impressive thing, deserving of more consideration than can be given here. Many groups were part of the campaign, spearheaded in the Senate by Democrat William Proxmire of Wisconsin (with whose office the C.L.A.S.B. maintained direct contact). The Coalition Against the SST, the Sierra Club, the Friends of the Earth, Common Cause, and the Environmental Defense Fund were active along with the League.

The SST is an environmental issue primarily because it would create a noisy and damaging sonic boom on the surface of the earth along the entire path of its supersonic flight; there are also other deleterious effects. But it is unlikely that much progress would have been made by the anti-SST lobby had not the economic feasibility of the plane been severely questioned by fifteen eminent establishment economists.[45] The president strongly backed the SST.[46] But unlike the ABM, there were economic criteria of profits and losses, for government as well as airlines and manufacturers, that could be applied to the SST.

Great anti-SST strength was shown in late 1970 when the Senate rejected federal funding for the Boeing commercial SST. But the Senate leadership appointed a pro-SST majority to the conference committee. In March 1971, both the Senate and the House of Representatives rejected funding for the SST. While these votes did not exclude construction of the SST through means such as private financing, nevertheless the congressional outcome was a legislative victory to which the C.L.A.S.B. contributed.

A large number of C.L.A.S.B. members have sufficient independent means to permit the organization to operate without membership dues. It is, of course, a lobby —not tax exempt and without foundation financing.[47] C.L.A.S.B. lobbying relied on cards, letters, and telegrams to congressmen by members and friends. Shurcliffe wrote a privately printed SST and Sonic Boom Handbook, later printed commercially with sales of 100,000 copies.[48] The league's news releases were used by many papers across the country. Indeed, a C.L.A.S.B. boom-path map of the North Atlantic Ocean was released by anti-SST Senator Gaylord Nelson (Democrat of Wisconsin) at a news conference, and thus was shown on national television. Shurcliffe and other active members of the league developed a great deal of technical information about the SST, an activity that brought them into some conflict with such institutions as the National Academy of Sciences and the Federal Aviation Administration.

Finally, among moderate groups there is Physicians for Social Responsibility. Formed in 1961, the P.S.R. has been centered in Boston and headed largely by distinguished members of the Harvard Medical School faculty (cardiologist Richard Feinbloom is chairman). There are chapters in several other cities, however. P.S.R. resembles S.S.R.S. in focusing on ethical problems faced by professionals. Aside from representing a collection of some 900 medical professionals concerned about their social responsibilities, the organization emphasizes research into the draft of doctors, chemical and biological warfare, and similar problems. The group's activities are financed on a modest annual budget of several thousand dollars, and it is a political interest group in that its critical publications and congressional testimony have been aimed at influencing public, professional, and legislative opinion.

The six groups discussed in this section are the major moderate political interest groups of science. There are some small new groups about which we have been able thus far to learn little—groups like Dentists for Peace. But what little we have been able to learn suggests that the six groups discussed here are the only ones that have gotten off the political ground to date, admittedly an impressionistic and common-sense judgment on our part. Recall also that local activity is not our concern here; there has been an increase in the past few years of local activity on environmental issues by a variety of types of professionals and students, with concerns similar to those of S.I.P.I. and C.L.A.S.B.

The Moderate Political Interest Groups

All of the moderate political interest groups exhibit a typically American pragmatism. They do not carry ideological frameworks with them into the political arena. This may be a strength in allowing them to be active, attacking specific policy problems head-on. This may be a weakness too, insofar as a pragmatic emphasis precludes political analysis and leads to a public attack on merely symptomatic problems of the American political system. We have criticized the pragmatism and optimism of the C.L.W. and F.A.S. at some length in our survey of their activities.

Every one of the six groups engages in efforts to inform public opinion on issues with a technoscientific component. Even the S.S.R.S. has done this, as in its campaign against the Vietnam war. The mass media are paying increasing attention to legislative and private critics of governmental policies, and this is due in part to the vigorous efforts of various critical interest groups—scientist and nonscientist—themselves. Yet these undifferentiated appeals to "the people" or "the public" presuppose the classical public of liberal democracy. Surely the number of persons approximately meeting these citizenship criteria is relatively small.[49] Primary social responsibility for this state of affairs does not lie with the people but with powerful institutions that perpetually propagandize them for their own purposes, from soap manufacturers to the president. In addition, elections constantly encourage citizen cynicism, since promises made during the campaign for the most distinguished office in the land are broken with such predictable regularity. Consequently, only people with ample spare time, and an ample stock of illusions about the responsiveness of the American political process, can follow specific issues closely and, more importantly, respond to appeals of scientist groups by communicating with political officeholders. Moreover, even if there is evidence of public concern, there remains the question of what congressmen do about it besides answering the letters sent to them. Political scientists have not demonstrated that letters from individual citizens generally have much effect on basic policies. These problems of public relations work constitute limitations of the appeal to a generalized public and point to the alternative of constructing political coalitions based on identifiable constellations of socioeconomic interests.

In addition to public education efforts addressed to a generalized public, the moderate groups all concentrate on education of members of Congress. In these efforts all the groups try to demonstrate the correctness of their positions and the validity of their information. Only the C. L. W. tries to supplement its informational liaisons with cash contributions. The limitations of this approach, too, are all too evident. The responsiveness of congressmen to organized groups and interests in their localities and states, accentuated by the seniority and committee system in the Congress, constitutes a political dynamic with which the scientist groups are ill equipped to cope. Furthermore, the center of foreign policy decisions is without the Congress, in the executive branch. The effort to educate congressmen has, of course, made some headway with some legislators around some issues. In general, direct efforts to shape congressional opinion have been a hallmark of the moderate style since the postwar scientists' movement. Moreover, the emphasis on the legislative arena does separate the moderate groups from the radical groups.

Finally, all of the moderate groups are headed by professionally established persons, mostly from recognized elite universities. Only S. I. P. I. receives significant governmental and foundation support, but the other groups benefit to a greater or lesser degree from wealthy members and/or benefactors. The elite cast of the moderate groups also distinguishes them from the radical groups.

THE NEW RADICALISM

The Old Radicalism

The radicalism that grew anew among American scientists in the 1960s was discontinuous with, but nevertheless foreshadowed by, an older radicalism. Although we began our discussion of the moderate style by describing the early scientist protests around atomic bomb and nuclear energy policy, the mid-1940s were not the historical watershed for the increase in political awareness among American scientists. Rather, the late 1930s were.

A now little-known group, the American Association of Scientific Workers, was formed in 1938. On the one side were the threats of persistent economic stagnation, and then of rising fascism abroad; on the other, the influence of the American Communist Party (particularly in intellectual circles) and the appeal for many of the Soviet Union (at least until its pact with Germany in 1939).[50] An A.A.S.W. notice in Science magazine observed that "scientific workers...as a group have virtually no control over the applications of science," and announced that the first purpose of the new group was to "bring scientific workers together to promote an understanding of the relationship between science and social problems."[51] One means of accomplishing A.A.S.W. purposes would be:

By giving expression to the collective opinions of scientific workers, and by taking appropriate action either individually, jointly, or through collaboration with existing organizations.[52]

The A.A.S.W. idea of collective organization of scientists, seen as workers, to assert greater control over the social uses of their labor was a bold conception, and certainly differed from the later rationale of the moderate scientists' movement, that scientists have a special responsibility to provide information to the public and lawmakers. Although the political outlook of the A.A.S.W. resembled that of the new political groups of professionals that developed in the 1960s, there was no continuity (except for a few individuals active in both periods); the A.A.S.W. virtually disappeared from the political scene after World War II.

The several moderate groups discussed earlier are open to scientific workers of any rank (and the S.S.R.S. even sees itself as an organization of "scientific workers"). Yet the emphasis on collective organization of workers does set the A.A.S.W. off from any of these. In congratulating the A.A.S.W. on its formation, the head of the Association of Scientific Workers (Great Britain) wrote:

It is essential for scientific workers to organize both to protect their own economic and professional status and to work for the better organization and application of science for the benefit of the community.[53]

The A.A.S.W. never developed the trade-union aspect of its political outlook. In general, its ideas were radical but its action were not. It was principally a discussion forum. Its actions as a group were primarily aimed at influencing public opinion or governmental opinion. Its leaders were all university scientists.[54] Its most significant act was the submission of a neutralist petition, signed by 500 scientists, to President Roosevelt in 1940.[55] With the circulation of the petition, A.A.S.W. ideas bore a remarkable resemblance to those of the new radicalism, complete with opposition to military and industrial exploitation of technoscientific skills. (Even consumerism was foreshadowed when the head of the Consumers Union spoke to the A.A.S.W. on "The Relation of the Consumer Movement to Scientific Groups.")[56]

Just how far left was the A.A.S.W.? The only political scientist who has paid much attention to it, Donald Strickland, came across an informant (in the course of interviews on the scientists' movement of the 1940s) who had been a Communist Party official charged with organizing intellectuals in "a midwest urban area" and who stated that the A.A.S.W. had been organized by the party.[57] There clearly were some Communists in the group. Its neutralist petition was drawn up after the German-Soviet pact of 1939. Dissident members of the particularly active Cambridge chapter accused the local branch of being Communist dominated, and several members resigned.[58] After the war the A.A.S.W. joined with several communist-oriented counterpart groups through the World Federation of Scientific Workers (W.F.S.W.).[59] The A.A.S.W. definitely had a left-wing reputation.

Yet all this evidence is tenative and circumstantial, for a study of the A.A.S.W., as such, has never been done.[60] The Communist Party of the U.S. had a very diffuse political line—beginning with the Popular Front (a united front against fascism which entailed cooperation and collaboration with anticommunist groups that opposed fascism) of the 1930s and culminating in the actual dissolution of the U.S. party in the closing months of World War II—so it is not clear whether the many associations and groups in which the communists worked were dominated by them, or dominated the communists. In any event, the A.A.S.W. had moderate as well as liberal and left members.[61]

The important point about the A.A.S.W. is not that of its exact relationship to the Communist Party. The point is, rather, that the group represented an incipient radical politics which the postwar scientists' movement rejected. As Strickland puts it, "a deliberate decision was made to dissociate the two movements, and the A.A.S.W. sympathizers were effectively quarantined...."[62] F.A.S. was anticommunist from it organizational inception. Its first secretary was dismissed because she was a member of the Communist Party. The chairman of the F.A.S. chapter at Fort Monmouth,

New Jersey, resigned because he questioned the loyalties and sympathies of the members. A radical group in the New York City F.A.S. chapter was barely defeated for leadership of that important body. All suggestions that the F.A.S. ally with either the A.A.S.W. or the W.F.S.W. were rejected.

The A.A.S.W. lingered on for several years, and we do not know precisely when it expired. The first organized radicalism in American science may have been the victim of its own idecisiveness and failure to build any type of action program. It was caught between the anticommunist crusade of the American government on one side and on the other the politics of a Communist Party dependent on a foreign government. As for the moderate political movement, it can now be seen that it was not born in a political vacuum; rather a specific rejection of a radical policy outlook took place at its inception. Such radicalism as existed among American scientists was not organized through a national political interest group again until the mid-1960s.

The Medical Committee for Human Rights[63]

In 1964, young blacks spearheaded protests and revolts in New York, Philadelphia, Chicago, Jacksonville, and other cities. Since then, no year has gone by without manifestations of a black rebellion. In 1965, the war in Vietnam was escalated, and 30,000 protestors marched in Washington under Students for a Democratic Society auspices. Since then, no year has gone by without organized antiwar protests. The conditions and policies that triggered these rebellions and protests persist, and so the actions recur. They are symptoms of a deepening crisis in the American political and social system, part of which involves the perceptible erosion of governmental authority during that past half-dozen years. One response of some scientists has been to form new types of organizations designed to take more direct action against per- ceived injustices—groups that are not predicated upon a confidence that government, prodded by the informed lobbying campaign, will respond to widespread or intensely felt public opinions and needs.

The Medical Committee for Human Rights was originally formed in 1964 to provide medical care during the civil rights demonstrations and marches. It continued to work with Martin Luther King and the Southern Christian Leadership Conference and also provided medical care at antiwar and other protests. Gradually it developed into a broader and more radical national organization and, in 1971, had twenty-four local chapters. An increasing number of young doctors, nurses, and medical students are openly discontented with the political economy of medical care, and with their support M.C.H.R. has grown to about 6,000 members. The M.C.H.R. is also open to all medical workers and sees all, scientists and nonscientists, professionals and workers, as "health workers."

The M.C.H.R. spnsored a demonstration at the 1970 convention of the A.M.A. at which they set forth the following demands:

End racism in the health care delivery system.
End oppression of women by and in the system.
End war-collaboration by the health industry.
Socialize the health care system. [64]

The policy outlook of the M.C.H.R. is clearly radical. At its center is an anticapitalist attack on the economic system, at least insofar as health care is concerned. On the one side, it sees health as a human right. On the other, it sees a complex of vested interests around health care—the drug industry, the fee-for-service independent doctors, the hospital supply industry, the medical electronics industry, nursing homes, and proprietary hospitals—all motivated primarily by private profit. In the view of M.C.H.R., developments in the health field have not changed the character of the system at all; government assistance is now welcomed by the interests:

It was Medicare that transformed the old bogeyman of government "interference" into a Santa Claus for the health industry. [65]

Nor, according to this group, are any innovations being contemplated by government expected to change the profitability of the health industry for capitalists, its utter inadequacy to the poor, and its increasing cost to middle-income groups. Politically powerful capitalists now do want national health insurance. We have a "two-class medical system," and, in the view of two observers:

National health insurance would be a mechansim for funneling money out of the pockets of workers and taxpayers into the hands of the people who now run...the system—the doctors, the hospital administrators, and the medical-industrial complex that fattens off people's illness. [66]

The M.C.H.R. distinguishes sharply between liberal critics of the laissez-faire attitude of the politically powerful A.M.A. (which successfully pressured President Nixon to withdraw an offer of an assistant secretarial appointment in the Department of Health, Education, and Welfare from Dr. John Knowles) and its own perspective:

To speak of Dr. Knowles as a better candidate than the A.M.A.'s antediluvian choice is to misunderstand the nature of the capitalistic health economics.... Dr. Knowles is part of the progressive wing of the capitalistic class. [67]

These excerpts make clear the general policy outlook of the look of the M.C.H.R. Its economic critique is, relatively, much sharper and more explicit than that of the old A.A.S.W. Discussion of political and economic issues represent an important aspect of M.C.H.R. activity. This sets the group off from the moderate groups which, with their pragmatic approach, place much less emphasis on analysis of American politics in their activities. Beyond the central anticapitalism of the policy outlook, there are a number of complementary analyses and positions that cannot be explored in the space available here. Briefly, they deal with the curriculum and control of medical schools; control of and practices of hospitals; the relation of imperial war to health at home (and abroad); racism and sexism; and the key political question (as yet unresolved within M.C.H.R.) of the value of beginning efforts to shape federal legislation. In contrast to its theoretical analysis pointing to an increasingly monopolistic and state-supported exploitive central economy of health, M.C.H.R. political strategy emphasizes local action. Most members expend their political energy in locally chosen chapter actions, while a small group runs M.C.H.R.'s national business. The principal activities of local chapters revolve around free clinics. By early 1971, there were about 125 such clinics providing free medical or dental, drug, or psychological services.[68] The largest number of clinics has been set up by community groups—Black Panthers or Young Lords or groups of neighborhood residents or local medical students and professionals concerned about the fate of hippies near campuses. (However, about a quarter of the clinics were set up by local governments or businessmen to serve local youths.)

Attempts by M.C.H.R. chapters to set up their own clinics have been very rare (although some, particularly at the level of pregnancy detection or the drug assistance "rap" center, have been begun). Usually M.C.H.R. chapters respond to calls for assistance from others beginning clinics. They may provide medical volunteers; train nonmedical personnel in screening patients, performing routine tests, and taking their histories; or otherwise help in organizing the clinic. Some M.C.H.R. chapters have contributed money to clinics (typically plagued by financial problems) and generated favorable media coverage of them (when threatened by hostile public health officials, as in Chicago).

Some M.C.H.R. chapters have supported efforts of unions like "Local 1199" to organize campus or health workers. Other chapters have done studies of aspects of health or health care to root out information not available from official authorities. The national M.C.H.R. has suggested a program whereby local chapters can challenge the accreditation of hospitals, but it is not determined or coordinated nationally. In the local chapters there is a conscious emphasis on collective political discussion and decision. (This is also true of the clinics, many of which emphasize antiracism and antielitism—in short, are political.)

The limitations of M.C.H.R. activism spring from the centralized character of the economic problems of health. Assistance to the free clinic movement can provide real services to some people in need, but the movement clearly does not change the basic character and impact of the health economy. However, M.C.H.R. decided at its April 1971 convention to supplement local activism with a nationally coordinated campaign for a socialized, yet locally controlled, national health system.

The M.C.H.R. also publishes reports of chapter activities; news about struggles in the health field, as for example, over community control of hospitals, and political analysis of the economy of health. Its publications are a newspaper, Health Rights News, and a magazine, The Body Politic.[69]

The national M.C.H.R. has occasionally prodded the federal government, as by writing the commissioner of the Food and Drug Administration about the hazards of commercially sold chemical Mace. The principal activism of M.C.H.R. as a national group has centered on the A.M.A. conventions. Since 1965, M.C.H.R. demonstrators have confronted the A.M.A. In 1969, many seized the stage; speeches and clenched fist salutes were given, amidst much hostility from the delegates and generous national press and television coverage. In 1970, M.C.H.R. sponsored a "People's Health Convention" at the A.M.A. meeting, in which several other organizations, including women's liberation, consumer, and welfare rights groups, participated.

It is as yet unclear how important the nationally planned health care campaign will become vis-à-vis the local activism that remains the main channel of M.C.H.R. energies thus far. Advancing the national health care plan does entail attempting to influence Congress in a not untraditional fashion (an endeavor that could transform M.C.H.R. into a moderate political interest group), rather than an emphasis on linking up with other groups in some sort of general political struggle for socialism.

Scientists and Engineers for Social and Political Action

Originally called "Scientists for Social and Political Action," S.E.S.P.A. was formed by physicists at a meeting of the American Physical Society early in 1969. Its early activities were aimed at bringing politics, particularly the politics of the ABM, into the proceedings of professional meetings of physicists.[70] S.E.S.P.A. collaborated with the F.A.S. in anti-ABM activities and, at an April 1969 American Physical Society national meeting, polled 1,200 members, determining that 76% were opposed to the Nixon ABM proposal.[71] S.E.S.P.A. was also active, in a supplementary fashion, in contacting senators and representatives to persuade them to oppose the ABM.

At its formation, S.E.S.P.A. was not radical in conception. It announced its organization as "a forum where all concerned scientists, and especially younger members of the profession, may explore the questions, "Why are we scientists? For whose benefit do we work? What is the full measure of our moral and social responsibility?"[72] But following the Senate's defeat of the anti-ABM forces, S.E.S.P.A. developed a broader and more radical policy outlook (and assumed its present name, enrolling many nonphysicists). Scientists were seen as "workers" and the slogan "Science for the People" was adopted:

We scientists are workers. Our only hope in preventing further misuse of science is to join with all other workers to bring about a radical change in the thinking, goals, and economic structure of this country.[73]

Activism was advocated, and the liberal argument that radical action provokes backlash and repression was explicitly rejected. The best defense against repression was held to be collective criticism and action; "When they threaten our jobs if we speak out, we must speak out."[74] As to unemployment among professionals, no solution was seen within the framework of capitalism:

The only hope for the great majority of unemployed scientists and engineers is not in individual adaptation, but in structural change of the conditions which determine their unemployment perspectives. Defense employment and useful civilian employment... are structually asymmetric, and one cannot be substituted for the other...without a basic change of the system.[75]

S.E.S.P.A. holds that individual solutions and individualism cannot help techno-scientific workers or any workers and, indeed, are the key to the continuation of capitalist exploitation:

Fragmentation among working people is part and parcel of the American ideology and it serves capitalism well. Whites are set against blacks, men against women, white-collar workers against blue-collar workers. Disunity in the guise of individualism is elevated to a virtue thus quite effectively preventing concerted action of workers in their common interest.[76]

The problem of misuse of science (particularly through military exploitation), like the problem of economic condition of scientific workers, is thus soluble only through what one S.E.S.P.A. member called "a total reorganization of the society on socialistic lines."[77] The (unscheduled) S.E.S.P.A. speech at the 1970 A.A.A.S. meeting, given just before N.A.S. President Philip Handler's major address, stated in part:

Philip Handler is going to talk to you...about...how the scientific community can help prop up the ruling class' corporate profit...We're here in the interest of those people who are not interested in rationalizing their rule, but in destroying it....
What is needed now is not liberal reform...but a radical attack....Scientific workers must develop ways to put their skills at the service of the people....
You still have the opportunity to work constructively with the movement for revolutionary change.... But if no other solution is available, we will be out in the streets,... doing everything we can to tear this racist, imperialist system to shreds....[78]

Clearly, the policy outlook of S.E.S.P.A., like that of the M.C.H.R., is radical. But also, like the M.C.H.R., S.E.S.P.A.'s political strategy emphasizes local direct action, despite the theoretical critique of capitalism as a national system of domination with a ruling class. Most S.E.S.P.A members expend their political energy in locally chosen chapter actions, which thus far have not been subjected to national approval. In the San Francisco area, from which came the initiative for forming S.E.S.P.A., the Berkeley chapter has demonstrated against the University of California's Livermore Laboratory, a military research facility, and has organized local commemorative protests marking the first anniversary of the March 4, 1969, work stoppage at MIT (and other schools) and the twenty-fifth anniversary of the dropping of the atomic bomb on Japan. In the Boston area, to which the center of S.E.S.P.A. political activity appears to have shifted and where the organ Science for the People is published, members have publicized military research and have held joint meetings with other similar groups that have sprung up.

The major national S.E.S.P.A. actions have centered on the meetings of the A.A.A.S. The first organized challenge to speakers took place at the 1969 meetings in Boston. In the 1970 Chicago meetings, there were fewer challenges of speakers and format, but the confrontations that did take place were chosen more carefully. During a talk, Dr. Edward Teller (well-known as the "father of the H-bomb") was criticized with on-stage placards and pantomime and later presented the "Dr. Strangelove" award, which he refused. At one of the confronted seminars on "Crime, Violence and Social Control," S.E.S.P.A. members succeeded in changing the structure of the panel, making comments critical of the fundamental assumptions of some panelists and eliciting extensive discussion from and among the audience. S.E.S.P.A. also drew up an indictment of the new A.A.A.S. president, Atomic Energy Commissioner Glenn T. Seaborg, which Seaborg evaded by leaving the panel at which he was to have spoken; he was charged with "the crime of science against the people."[79] Additional activities at the meeting included guerilla theater, political workshops, and circulation of a pledge.[80]

S.E.S.P.A. leaders say that the group is a "nonorganization—a group with no officers and no constraints on membership," for

If groups are to struggle against nonparticipatory, undemocratic structure, it is necessary that they don't replicate such structure in their own organizing.[81]

Lack of formal organization does set S.E.S.P.A. apart from the M.C.H.R. Even the national action at the A.A.A.S. was organized principally by the Chicago S.E.S.P.A., with assistance from the University of Chicago New University Conference

"People's Science Collective." There are S.E.S.P.A. study groups, and, as in
M.C.H.R., there is an organizational principle of collective discussion, decision,
and self-criticism. Organization work is done on a volunteer basis, and the major
cost—publication of the magazine—is offset by magazine and button sales and individual contributions. We do not know how many members there are; the number is
probably between S.E.S.P.A.'s minimum estimate of 2,500 and the number of issues
of the August 1970 issue of the magazine distributed (5,000). Some of the most active
S.E.S.P.A. members have suffered harassment, including docking of pay and outright firings from jobs; one, William Davidon, was cited on January 12, 1971 in the
"plot" the Federal Bureau of Investigation has claimed was formed to kidnap national
security advisor Henry Kissinger (Davidon is also an S.S.R.S. leader).

Some Characteristics of Group Leadership

In our analysis of the moderate political interest groups we devoted some attention to
the leaderships of those groups. Such profiles of leadership are not in themselves as
useful as firsthand in-depth studies of the political dynamic of internal decision making.
Hopefully, however, they contributed to our first approximation of a descriptive analysis of the politics of the moderate scientist groups. Analysis of the leadership of the
radical groups is more difficult because data on the leaders have been more difficult
to obtain, and in the case of S.E.S.P.A. there is no formal structure that can even
be studied.

Theoretically all S.E.S.P.A. members are "leaders." But if the group has avoided
institutionalization and if its internal affairs are democratically decentralized, some
members are nevertheless more active than others over time. There is no such thing
as a political group without leaders, even if leadership is informal, democratically
responsive, and/or frequently rotated. In the S.E.S.P.A. case, additionally, Berkeley
University physicist Charles Schwartz and Boston industrial physicist Herb Fox have
been notably active over the past two years; were there officers, they would surely be
among them. M.C.H.R. does have a formal national structure and formal internal
elections; its chairman is Quentin Young, a Chicago medical doctor, and its vice-
chairman is Felicia Hance, a woman doctor from San Francisco.

For purposes of this analysis, S.E.S.P.A. "leaders" were held to be the four
founders; members mentioned as active in the New York Times, Science, and Science
for the People; members of the editorial collective; and local chapter contacts. Of
seventeen S.E.S.P.A. activists for whom we have data, 82% are in natural science
fields, principally physics. About a third of the S.E.S.P.A. activists are located in

Table 7. Political Groups, Ranked by Percent of Leaders Listed in Standard References[a]

Rank	Group	Percent Listed	Number of "Leaders"
1	F. A. S.	90	29
2	C. L. W.	82	17
3	S. E. S. P. A.	29	31
4	M. C. H. R.	23	26

[a]References used are American Men of Science and American Medical Directory. Data are for leaders as of 1971, although the four 1969 founders of S.E.S.P.A. are included as 1971 leaders. M.C.H.R. has a list of sponsors who lend their names to the group, but these are not included as leaders. Local M.C.H.R. chapter leaders are not included either.

Massachusetts. Two S.E.S.P.A. leaders are also F.A.S. leaders. In general, the leadership of the two radical groups is less professionally established than the leadership of the F.A.S. and C.L.W., as can be seen from Table 7.

Six of the 26 members of the M.C.H.R. national executive committee are listed medical doctors. Although the committee differs considerably from the 1970 group —of which 63% were listed doctors—and now functions on a collective basis, the M.C.H.R. is much more formally structured than S.E.S.P.A. There is a treasurer, a Philadelphia registered nurse, and a paid staff headed by a former union organizer. Although scattered data about the positions and careers of M.C.H.R. leaders make it clear that they, like S.E.S.P.A. leaders, are a much younger group than the moderate leaderships, we do not have an average age figure.

The Radical Style

Although the two radical groups just discussed are the principal radical interest groups of science, the much smaller Computer People for Peace (C. P. P.) in many ways mirrors their concerns. When founded early in 1968 as "Computer Professionals for Peace," the C. P. P. was—as M.C.H.R. and S.E.S. P.A. at first were—a liberal antiwar group. As some C. P. P. members challenged professional societies and corporations through panel disruption and demonstration, the group developed the commitment to direct action that is the hallmark of the radical style. This commitment did not exclude other forms of activity, such as fund raising for a New York computer programmer (Clark Squire) accused in the "Black Panther 21" conspiracy trial of 1970-1971; but the commitment to direct action did go hand in hand with the

development of the socialist ideological outlook that also characterizes contemporary
radicalism. The "leaders" of the C. P. P. —its most active members—are a young
group of fewer than twenty, principally members of the informal "Steering Committee" based in New York City. More of these activists are in industry jobs than is
the case in S. E. S. P. A. and M. C. H. R., and the firings they have experienced have
been somewhat more frequent (relative to the small total membership of about two
hundred) than with the other radical groups. C. P. P. publishes the newsletter
Interrupt, and its slogan now is "Technology for the People!"

There are no other radical membership groups.[82] Of course, some militant radicalism is outside of the scope of our overview in that it has been only local in scope.
The university revolt is nationwide, but largely uncoordinated nationally. For example,
MIT has been the scene of particularly protracted local activism spearheaded by
science students, one result of which has been the divestiture by the Institute of a
major military laboratory.[83] Other radicalism has occurred with the framework of
nonscientist groups—New University Conference, Young Socialist Alliance/Student
Mobilization Committee, Students for a Democratic Society, and the Progressive
Labor Party—within which science teachers and students have been active.

So far as the radical scientist political interest groups are concerned, several
conclusions emerge.

First, as to policy outlook: the old A. A. S. W. idea that scientists should organize
and assert greater control over the application of their labor pales when compared
with the explicit anticapitalism of the new radicalism. Moderate and radical groups
share a dissatisfaction with public policy toward the exploitation of science, yet it
would be difficult to overemphasize the gulf in political attitudes separating the radical groups from the moderate ones. The radicals' attack on capitalism as such and
as involving racism, exploitation of women, and elitist individualism is, thus far,
unremitting. Emphasis on the value of linking up with other workers and other radical
movements is increasing (though in practice these bridges have yet to be built). Much
more attention is devoted to the analysis of the political economy of public policy by
the radical groups than by the moderate ones; thus the radical policy outlook is not
only critical on the economic as well as the political plane but is also more developed
(at least quantitatively). The policy outlook of the radical groups is characterized, not
by pragmatism, but by a relatively developed anticapitalist theory.

Second, as to the political strategy: the radical emphasis on direct action contrasts
with the moderate emphasis on the legislative arena. Moderate politics presupposes
the validity of the traditional democratic policy process. Radical politics consciously

ignores that process, denying its democratic validity. Many political scientists might not even view confrontations, disruptions, demonstrations, demands, sit-ins, and community action as political. Certainly, direct action in the workplace, neighborhood, hospital, school, or professional meeting represents a quite different form of political action from that employed by the moderate groups. The constant practice of collective political discussion in connection with action projects is also a hallmark of the radical style. National organization is weak and organizational structure decentralized. S.E.S.P.A. is not even formally constituted. Only one of the moderate groups, S.I.P.I., has the emphasis on local chapter action that typified the radical groups. Because of the diffuse character of radical political efforts, it is difficult to devise a fair standard for judging their success to date. If it is the abolition of the capitalist economic framework within which technology and medicine are applied, then obviously little progress has been made.

There are some similarities between the radical and the moderate style. To some extent the radical groups employ a generalized public, as do the moderate groups. There is an emphasis on what "the people" want which is not supported by the kind of extensive analysis that underpins the anticapitalism of the policy outlook.

There are no obvious disciplinary differences distinguishing the leaderships of the radical groups from the moderate ones. Physicists are prominent in S.E.S.P.A. but also in the leadership of moderate groups (especially F.A.S.). Medical doctors are active in M.C.H.R. but also in moderate group leaderships (notably P.S.R.). Further analysis might suggest some kinds of associations between discipline and political attitude, but at first glance our overview here does not replicate the kinds of disciplinary contrasts found in the author's MIT study and suggested by other students of scientists in public affairs. Various types of scientists, and some social scientists, are scattered among the leaderships of several of the groups. There is virtually no one from the humanities in the radical groups or the political interest groups of science generally.

CONCLUSION: PROBLEMS OF SCIENTIST POLITICS

Before concluding this survey we remind the reader that we are dealing with only one aspect of scientist politics, albeit an important one. We have not considered that small number of scientists who have constituted an advisory elite and have viewed their access to the presidency and other executive agencies as an important means of possible policy influence. There is no space here for an assessment of the content and degree of advisory influence. The following comments are therefore directed not at institutional

policy participation, whose limits we have not been able to explore here, but at political interest group activity.

In 1971, perhaps as many as 35, 000 persons were members of scientist political
interest groups. [84] The number has increased rapidly in the past decade as new moderate and radical groups have been formed and have grown. We do not know how many
members are scientists (where doctors, engineers, and students are included in our
definition), how many are technical workers, and how many are not in technoscientific skill situations at all; the probablility is that most, thus far, are scientists and
science students. The absolute number of members is large and increasing but must
be kept in perspective. The total number of members is but a small fraction of the
membership of the scientific occupations and is divided among four major and several
minor groups. There is no single principal political group, either radical union or
moderate lobby, that has emerged in the scientific occupations. Still, the increasing
number of individuals challenging major and minor public and politico-economic policies—albeit in an as yet uncoordinated way through a variety of groups—cannot be
ignored.

In this overview we have not dealt with the normative question of the validity of the
goals of the scientist groups. We have, however, dealt with the analytical question of
the utility of the present political strategies for attaining the stated group goals. The
only moderate group that at this writing appears to have contributed to substantial policy change is the C. L. A. S. B., which had a specific and limited objective, was part
of a much broader lobbying coalition, and fought against a project that increasingly
appeared economically unfeasible. Insofar as the moderate groups have a political
analysis, it tends to ignore the executive branch of the national government and points
to public opinion and the legislative arena, political strategy is consistent with these
foci. In this way the question of what interests determine the direction of presidential
policy, which is at the political center of the American system, is sidestepped in theory
and practice. [85] This omission is especially striking where arms control is the issue;
any relation of arms policy to broader aspects of American foreign policy is substantially ignored, as is the obviously central role of the executive in the foreign and military policy-making process.

The political importance of the executive center of the national government is implicitly acknowledged in the policy outlook of the radical groups. The radical analysis
sees an industrial complex or capitalist ruling class that the executive branch either
represents or is part of. But while this analysis points squarely to an executive center
of irresponsible power, the political strategy focuses on local action. How local action,

not to mention local action by a nonorganization, can combat a ruling elite is, to say the least, not self-evident. The direct actions of the radical groups are not analytically or actually linked to the national fact of an executive of the state that enforces existing policies and has armed force at its disposal.

Thus, it appears that the executive branch of the federal government and the interests it represents are unlikely to be moved substantially by the scientist political interest groups. The presidential center seems oddly secure from the pressures of moderate and radical groups alike. Both sets of scientists seem rather far from actually being able to accomplish their aims.

NOTES

1. For helpful comments on an early draft of this chapter, the author thanks Edwin Layton and Kenneth Zapp of Case Western Reserve University.

2. We are adapting George I. Blanksten's category "associational political groups," defined by him as "consciously organized associations which lie outside the formal structure of government and which nevertheless include the performance of political functions among their stated objectives." "Political Groups in Latin America," American Political Science Review, Vol. 53, March 1959, p. 108. Blanksten feels it redundant to modify "group" with the two adjectives "political" and "interest," but we do not.

3. In Scientists and National Policy Making, Robert Gilpin and Christopher Wright, eds. (New York: Columbia University Press, 1964), pp. 41-72. Wood's use of the term "scientist" is more restricted than ours, as is the case with most discussions of "scientists" and public affairs. The "skill group" concept he uses is based on the work of Harold Lasswell in Politics: Who Gets, What, When, How (Cleveland: The World Publishing Co., 1958).

4. Harold Lasswell projected "the further rise in influence of persons skilled in mathematics, physics, and chemistry," apparently displacing businessmen and politicians. Politics, p. 196.
   The economist Robert Heilbroner projected that "new elites" who "look to science as the vehicle of their expertise" will (some foreseeable day) be in a position to assert the primacy of their aims "over the aims of the existing order," apparently displacing the present ruling elite of American capitalism (and that of Soviet socialism too). See The Limits of American Capitalism (New York: Harper & Row, 1966), pp. 128-129.
   Though holding back from any projection of scientists in a role as the ruling elite, Robert Wood also anticipates a new top-level "influence" and "power" for the skill group. "Scientists and Politics: The Rise of an Apolitical Elite," in Scientists and National Policy Making, Gilpin and Wright, eds., p. 46.

5. Among recent studies on the subject are Michael Reagan, Science and the Federal Patron (New York: Oxford University Press, 1969), and Don K. Price, The Scientific Estate (Cambridge: Harvard University Press, 1965).
   A pioneering study of the federal scientist administrator is made by Eugene Uyeki and Frank B. Cliffe, Jr., "The Federal Scientist-Administrator," Science, Vol. 139, March 29, 1963, pp. 1267-1270. To our knowledge, this study has not been followed up

in published sources, although Melvin Bolster is presently engaged in a study of scientist administrators at the National Institutes of Health.

6. Among relevant studies are Carl William Fischer, "Scientists and Statesmen," Knowledge and Power, Sanford A. Lakoff, ed. (New York: The Free Press, 1966), pp. 315-358; Gilpin and Wright, eds., Scientists and National Policy Making; Daniel Greenberg, The Politics of Pure Science (New York: New American Library, 1968); and Frank von Hippel and Joel Primack, "The Politics of Technology" (Stanford, Calif.: Stanford Workshops on Social and Political Issues, September 1970).

7. For book-length treatments, see Oliver Garceau, The Political Life of the American Medical Association (Cambridge: Harvard University Press, 1941), and Elton Rayack, Professional Power and American Medicine (Cleveland: World Publishing Co., 1967).

8. Social scientists and members of the humanities faculty were excluded, and MIT has no medical school.

9. "The Political Attitudes of a Scientific Elite," Ph.D. dissertation, Department of Political Science, MIT (Cambridge: 1968). The respondents were guaranteed anonymity (as many were quoted), and as an additional safeguard the dissertation did not identify the locus of the study as MIT. With the passage of some years there seems little danger to the interests of the respondents in identifying MIT, although personal anonymity must be retained.

10. For some early speculation on this matter, see Richard L. Meier, "The Origin of the Scientific Species," Bulletin of the Atomic Scientists, Vol. 8, June 1951, pp. 169-173; Don K. Price, Government and Science (New York: Oxford University Press, 1953), especially p. 120; and C. P. Snow, The Two Cultures and the Scientific Revolution (New York: Cambridge University Press, 1961), especially p. 33.

Some studies have been done; much of the available evidence is summarized in Walter Hirsch, Scientists in American Society (New York: Random House, 1968), pp. 17-20. Recent studies not discussed by Hirsch include Everett C. Ladd, Jr., "Professors and Political Petitions," Science, Vol. 163, March 28, 1969, pp. 1425-1430; Henry A. Turner and Charles B. Spaulding, "Political Attitudes and Behavior of Selected Academically-affiliated Professional Groups," Polity, Vol. 1, Spring 1969, pp. 309-336; L. Vaughn Blankenship, "The Scientists as 'Apolitical' Man," unpublished manuscript, 1970; and Everett C. Ladd, Jr., and Seymour Martin Lipset, "Politics of Academic Natural Scientists and Engineers," Science, Vol. 176, June 9, 1972, pp. 1091-1100.

It seems clear from these studies that as a group engineers are more conservative than (other) natural scientists. The largest sample was that analyzed by Ladd and Lipset who, using a fairly conventional range of liberal-conservative issues and indices, found discipline to be the variable most useful in political differentiation among respondents. In their analysis—which had no "radical" category and which did not single out the politically active—physicists were most liberal, with increasing conservatism among faculty persons in medicine, mathematics, all biological sciences, chemistry, and all engineering, in that order.

11. Blanksten, "Political Groups in Latin America," p. 106.

12. For assistance in researching the second and third sections of this chapter, the author gratefully acknowledges the indispensable efforts of Joseph Leo.

13. The definitive historical account of the scientist movement of the 1940s is Alice Kimball Smith's book A Peril and A Hope (Chicago: University of Chicago Press, 1965). See also Donald A. Strickland, Scientists in Politics (Lafayette, Ind,: Purdue University Studies, 1968).

14. "The Franck Report," quoted in Smith, A Peril and A Hope, p. 567.

15. Ibid., p. 572.

16. "We were troubled...by the fact that except for the scientists and the Army there was no foreknowledge of the tremendous implications of the atom bomb," wrote a participant in the scientists' lobby (Aaron Novick, "A Plea for Atomic Freedom," New Republic, Vol. 114, March 25, 1946, p. 400). There was a "broad responsibility of scientists today" informing F.A.S. action, according to the organization's stated aims (F.A.S. files quoted in Smith, A Peril and A Hope, p. 236).

17. There is no single book dealing only with scientist political activism between the scientists' lobby and the mid-1960s. However, Robert Gilpin's book, American Scientists and Nuclear Weapons Policy (Princeton: Princeton University Press, 1962), deals with scientists in and out of government in the period. In addition, the collection of articles from the Bulletin of the Atomic Scientists edited by Morton Grodzins and Eugene Rabinowitch and entitled The Atomic Age (New York: Basic Books, 1963) reflects the political concerns of scientists in the period.

18. Primary information sources for this discussion are press reports of F.A.S. activity; the F.A.S. Newsletter; and communications with F.A.S. personnel.

19. For York's arms control philosophy, see his Race to Oblivion (New York: Simon and Schuster, 1970).

20. For Stone's arms control philosophy, see his Containing the Arms Race (Cambridge, Mass.: The MIT Press, 1966). Stone, whose personal efforts are the key to the continuing activity of the group, involved York in F.A.S. affairs.

21. Thomas A. Halsted, "Lobbying Against the ABM, 1967-1970," Bulletin of the Atomic Scientists, Vol. 27, April 1971, pp. 23-28.

22. For a news report on these activities, see Deborah Shapley, "FAS: Reviving Lobby Battles ABM, Scientists' Apathy," Science, Vol. 171, March 26, 1971, pp. 1224-1227.

23. "Federation of American Scientists," a 1970 F.A.S. brochure.

24. Jeremy J. Stone, "'Greening of America' Raises Questions for FAS," F.A.S. Newsletter, Vol. 23, December 1970, p. 2.

25. Principal sources for this discussion are press accounts of C.L.W. activities, C.L.W. literature, the C.L.W.'s Washington Bulletin, and communications with C.L.W. personnel. Secondary sources consulted include Leo Szilard, "Are We on the Road to War?," Bulletin of the Atomic Scientists, Vol. 18, April 1962, pp. 23-30; Elinor Langer, "Scientists in Politics: Council founded by Szilard Brings Cash and Sophistication to Lobbying," Science, Vol. 145, August 7, 1964, pp. 561-563. John McClaughry, "The Voice of the Dolphins," The Progressive, Vol. 29, April 1965, pp. 26-29; Richard L. Cohen, "Washington Pressures: Council for a Livable World," National Journal, February 14, 1970, pp. 360-364; and Halsted, "Lobbying Against the ABM, 1967-1970."

26. Speech at Amarillo, Texas, October 12, 1970. Quoted in a December 1970 C.L.W. letter to contributors. The data about Democrat McGovern's campaign are correct, even if the conclusion does not necessarily follow.

27. The leadership groups of the F.A.S. and C.L.W. are analyzed and compared in a later section of the chapter.

28. On the domination of the executive branch in the foreign-policy-making process, see G. William Domhoff, The Higher Circles (New York: Random House, 1970), pp. 111-155. It is important to note that some scientists have sought to influence arms policy by accepting advisory positions in the executive branch. Advisors of moderate policy outlook had access to the president at least from the 1957 reorganization of the President's Science Advisory Committee and creation of the post of personal presidential science advisor. An assessment of this form of political participation and its impact is without the scope of this survey. See David Nichols, "Scientists and Politics," a paper delivered at the 65th Annual Meeting of the American Political Science Association, New York, September 2-6, 1969, for an overview of the advisory process. For an analysis of the apparent recent decline in advisory influence, see Harvey Sapolsky, "An Exchange Model of the Science Advisory Process," a paper delivered at the 67th Annual Meeting of the American Political Science Association, Chicago, September 7-11, 1971.

29. The relationship between the broad foreign policy objectives of the United States and American arms policy needs to be explored by any group—whether of scientists, political scientists, politicians, or citizens—seriously interested in issues of war and peace. Major sources include Gar Alperovitz, Atomic Diplomacy: Hiroshima and and Potsdam (New York: Simon & Schuster, 1965), which shows that from its very invention as a weapon the atomic bomb was used by foreign policy decision makers in the context of their broad worldwide interests; Noam Chomsky, At War with Asia (New York: Random House, 1970); D. F. Fleming, The Cold War and Its Origins (Garden City, N.Y.: Doubleday, 1961); David Horowitz, The Free World Colossus (New York: Hill and Wang, 1965); Gabriel Kolko, The Politics of War (New York: Random House, 1968); Harry Magdoff, The Age of Imperialism (New York: Monthly Review Press, 1969); and William A. Williams, The Tragedy of American Diplomacy (New York: Dell, 1962).

30. Information about C.L.W. Vietnam war positions is derived from their literature of the 1967-1971 period and from communication with C.L.W. Director Thomas Halsted (who is not responsible for our analysis of C.L.W. politics).

31. From a description of the Saigon intellectuals' peace plan distributed by the C.L.W. in 1968.

32. The consistency of basic U.S. policy goals from administration to administration is brought out clearly in the definitive history of U.S. involvement, which stands quite well following the revelations in the "Pentagon Papers" of 1971. See George McT. Kahin and John Lewis, The United States in Vietnam (New York: Dial, 1969).

33. The age of one F.A.S. leader is not known.

34. Most of these data are based on information in American Men of Science, 11th ed., edited by the Jaques Cattell Press (New York: R.R. Bowker Co., 1968). For leaders not listed, information was obtained from group publications and communications directly to group leaders.

"Leaders" of the F.A.S. are taken to be its officers, its national council, and three chairmen of its membership, issues, and T.A.C.T.I.C. committees (the last deals with efforts to establish the advisory network for congressmen referred to in the discussion earlier). There is also a group of twenty sponsors who lend their names to the F.A.S., half of whom are well-known scientists like Jerome Wiesner, but these are not included as leaders in our analysis.

"Leaders" of the C.L.W. are its officers (including its counsel, known to be active in the group), its committee of scientists that chooses the board of directors, and the board of directors itself.

Data on the educational level of two C.L.W. leaders were unobtainable, but due to their occupations (staff director of the C.L.W. and documentary film producer), it was assumed that they hold no doctorates.

No weights were assigned in combined data on C.L.W. and F.A.S. leaders; the last column lists characteristics of all 43 politically active scientists.

Data are for leaders as of April 1971.

35. Primary sources for the groups discussed here are press reports of their activities, their literature or newsletters, and communications with organization personnel.

36. SIPI Report, Vol. 1, Fall 1970, p. 5.

37. Barry Commoner, Science and Survival (New York: The Viking Press, 1967), p. 131. The book is Commoner's own, but in its course he develops fully the S.I.P.I. rationale.

38. Data on S.I.P.I. leaders, and also on leaders of Society for Social Responsibility in Science and Citizens' League Against the Sonic Boom leaders, were developed as indicated in footnote 34 above, except that the twenty-fifth edition of American Medical Directory (Chicago: American Medical Association, 1969) was used to supplement American Men of Science when M.D.'s were not listed in the latter source.

39. Statement of Purposes adopted at S.S.R.S. meeting, 1960.

40. A wide array of scientists and students of science and public policy participated in the "Human Values in a Technological Society" meeting of the S.S.R.S. in Boston in 1970. Attendees ranged from the government advisor E. U. Condon to the historian of technology Melvin Kranzberg and the old Marxist Dirk Struik, and included people active in other groups like the F.A.S., Dentists for Peace, and Physicians for Social Responsibility.

41. The S.S.R.S. scientists were dispatched before a team headed by Matthew Meselson (incidentally a leader in the C.L.W. and F.A.S.) was sent on a preliminary investigation by the American Association for the Advancement of Science. For their report, see Gordon H. Orians and Egbert W. Pfeiffer, "Mission to Vietnam," Scientific Research, Vol. 4, June 9, 1969, and June 23, 1969.

42. The S.S.R.S. operating budget is under $10,000 annually.

43. Eighty-four percent of the Americans in the S.S.R.S. leadership are listed in American Men of Science. Of the leadership as a whole (assuming no doctoral degrees for three for whom educational data were lacking), 71% had doctoral degrees.

All data here are for S.S.R.S. leaders as of 1970.

44. The National Committee as such is less active than with the other groups, while much more work is done by local volunteers in Cambridge, Massachusetts, C.L.A.S.B. headquarters.

45. These included Milton Friedman, Arthur Okun, Paul Samuelson, and others similarly prominent. Samuelson saw the SST as "an economic and political disaster" (New York Times, September 16, 1970). The Anglo-French "Concorde" SST increasingly appeared a technological and, especially, economic disaster for those two countries.

46. He was supported by other members of the administration, including the Secretary of Transportation. Also, a well-heeled industrial lobby fought the anti-SST lobby with an intensifying public relations campaign.

47. C.L.A.S.B. took in about $60,000 in the first four years and was spending at the rate of $20,000 per year in March 1971.

48. William Shurcliffe, The SST and Sonic Boom Handbook (New York: Ballantine, 1970).

49. For a provocative discussion of this point, see C. Wright Mills, The Power Elite (New York: Oxford University Press, 1956), pp. 298-324.

50. See Irving Howe and Lewis A. Coser, The American Communist Party (New York: Frederick A. Praeger, 1962), pp. 273-318.

51. Science, Vol. 88, December 16, 1938, pp. 562-563.

52. Ibid., p. 563.

53. Sir F. Gowland Hopkins, quoted in Science, Vol. 89, January 20, 1939, pp. 58-59.
    The social-control-of-science movement began in England after World War I, and the British Association was always the largest and strongest of several national groups that eventually were formed. The British group became Marxist-oriented, and J. D. Bernal was especially active in it. On these developments, see Neal Wood's anti-communist treatment, Communism and British Intellectuals (New York: Columbia University Press, 1962), especially Chapters V and VI.

54. The president of the A.A.S.W. was the distinguished University of Chicago physiologist Anton J. Carlson. In a 1941 speech, he criticized not only German propaganda but also Britain's claim to be fighting for democracy, as "British big business has also fought for pure democracy in far-off India and Africa for quite a spell." He argued that a new "Age of Planning" was arriving:
    "It is the responsibility of the scientist to make the machinery of planning consistent with the structure of the democratic society. A democratic form of coordinated control must be developed in the transition from the political state to the social service state...."
    Quoted in Science, Vol. 93, February 14, 1941, pp. 159-160.

55. New York Times, May 20, 1940, p. 6.

56. Science, Vol. 91, March 8, 1940, p. 241.

57. Strickland, Scientists in Politics, p. 13.
    The following discussion of the A.A.S.W. and its relation to the F.A.S. relies primarily on Strickland unless otherwise noted.

There were other popular front groups, not dominated by scientists, which they joined; many individually were communists. This appeal of the left in scientific circles is vividly illustrated in accounts of J. Robert Oppenheimer's prewar milieu in Berkeley. See, for example, Philip M. Stern, The Oppenheimer Case: Security on Trial (New York: Harper & Row, 1969), pp. 15-28.

58. Among resigning scientists who are today well known were George Kistiakowsky and George Wald. The A.A.S.W.'s peace petition appears to have split the scientific community and to have restricted the growth of the group.

59. Smith, A Peril and A Hope, pp. 338, 525; Neal Wood, Communism and British Intellectuals, p. 165.

60. We are gathering information on the A.A.S.W. and hope to report elsewhere on this interesting episode in the political history of American scientists.

61. Smith, A Peril and A Hope, p. 126.

62. Strickland, Scientists in Politics, p. 15. There was an overlap of at least three leaders between the A.A.S.W. and the F.A.S.

63. Primary information sources for the discussions of the M.C.H.R. and Scientists and Engineers for Social and Political Action are their publications, press reports of their activities, attendance at their meetings, and communications with group personnel.

64. Quentin D. Young, "Welcome to Chicago," The Body Politic, Vol. 1, July-August 1970, p. 4.

65. Barbara and John Ehrenreich, "The Medical Industrial Complex," The Body Politic, p. 27.

66. John Ehrenreich and Oliver Fein, "National Health Insurance: The Great Leap Sideways," Social Policy, Vol. 1, January-February 1971, p. 9. The article lists beneficiaries of such insurance other than health care consumers. Fein is an M.C.H.R. leader, while Ehrenreich is on the staff of Health-PAC, a group discussed briefly later.

67. Richard Kunnes, "Off the System! (The U.S. Health Care Delivery System)," The Body Politic, p. 8.

68. Jerome L. Schwartz, "Free Health Clinics: What Are They?," Health Rights News, Vol. 4, January 1971, p. 5.

69. M.C.H.R. is federated with Psychologists for Social Action and the much smaller Physicians for Social Responsibility (discussed earlier) and Physicians Forum in an umbrella organization, the Council of Health Organizations. C.O.H.O. has about 10,000 members.

70. For a discussion of these early activities, see Charles Schwartz, "Professional Organization," in Martin Brown, ed., The Social Responsibility of the Scientist (New York: The Free Press, 1971), pp. 19-34.

71. Andrew Blank, "The Political Scientists" (unpublished Case Western Reserve University undergraduate term paper), December 1969.

72. Charles Schwartz, "Science: The Movement vs. the Establishment," The Nation, Vol. 210, June 22, 1970, p. 749.

73. SESPA News, undated 1970 issue.

74. Science for the People, Vol. 2, October 1970, p. 5.

75. Ibid., Vol. 2, December 1970, pp. 6, 8.

76. Ibid., Vol. 2, August 1970, p. 17.

77. Richard P. Novick, "Science Post-Sputnik Era: Epidemic Disillusionment," paper presented to the 137th meeting of the A.A.A.S., Chicago, December 27, 1970, p. 1.
    Novick is an M.D. and medical researcher and the scientist who presented Edward Teller with S.E.S.P.A.'s second annual "Dr. Strangelove" award for science in the service of warmakers.

78. Science for the People, Vol. 3, February 1971, pp. 6-7.

79. See the reproduced indictment in ibid., p. 12.

80. The pledge has been the subject of internal debate about its value as a radical tactic but is nevertheless being circulated under S.E.S.P.A. auspices. It reads, I pledge that I will not participate in war research or weapons production. I further pledge to counsel my students and urge my colleagues to do the same. M.C.H.R. has a similar pledge.

81. Ibid., p. 8.

82. A radical policy outlook is reflected by Physicians' Forum, a small group of doctors headquartered in New York since the 1930s. But, having the same reticence about activism that characterized the old A.A.S.W., the Physicians' Forum is not fully radical in our sense of the term. From 1965 to 1970 there existed a Student Health Organization founded as a counterinstitution agains the A.M.A.'s official Student American Medical Association. The S.H.O. was a network of local groups that may have been responsible for the development of a somewhat less conservative policy outlook on the part of the S.A.M.A. There is, finally, a Health Policy Advisory Center. Health-PAC is not a radical action group, but it does research health problems, provide advice, and put out a radically critical monthly Bulletin for groups and individuals critical of the political economy of health. Health-PAC is an incorporated group with a staff of ten, based in New York City.

83. See Dorthy Nelkin, The University and Military Research: Moral Politics at MIT (Ithaca, N.Y.: Cornell University Press, 1972).

84. An approximate sum of the memberships of the six moderate and two radical groups discussed here is 35,000. Overlapping memberships probably exist to some extent and would reduce this number; but on the other hand there are several small groups alluded to here whose membership has not been determined, and these would increase the estimate.

85. In dealing with "permanent" membership groups, we have been constrained to overlook at least one more transient type of organization which has attempted to influence the executive branch in rather direct sense, the quadrennial campaign organization, "Scientists and Engineers for Whomever-the-Democratic-Candidate-Happens-to-Be."

I. I. Rabi

It is very likely that the pattern of combining the efforts of a number of countries to form scientific laboratories or institutes will become even more important in the future than it has been in the past. As research becomes broader in scope and more exepensive, the costs go beyond what even a highly developed, middle-sized country such as France or England would find easy to support out of its own resources. Some kinds of talent are sufficiently rare so that only a few individuals exist in any single nation, and it requires a combination of scientists or experts from different countries to make progress.

In this study, three different international laboratories are examined through interviews, questionnaires, and other social science techniques, with a view to exploring the social and political aspects of such cooperation rather than the specifically technical values. The laboratories examined—CERN, ISPRA, and ESTEC—vary in quality and motivation. They are all in Europe and are all part of a basic political effort to find strength through cooperation. New institutions are in process of being organized at the present writing. Attempts are under way to establish an institute of computer science and one concerned with management. Both represent efforts on the part of certain European countries to close what has been loosely called the technological gap between the United States and Europe.

When CERN, the prototype of these efforts, was established to close the gap between Europe and the United States in the field of high energy physics, there was very little experience to guide the organization in setting up personnel policies and governance in general. Scientists were regarded as objective people, more or less alike in their habits of working and dominated by their scientific interests. Albert Teich's study shows that these assumptions were somewhat naïve—that the scientists take on to a great degree (certainly much more than one might have expected) the characteristic behavior patterns of their own nationalities. Beyond this, it appears that the scientists' response to the main purpose of the organizations which support them is an important factor in the effectiveness of their performance. Future attempts at organizing new international groups and installations will find the insights in this study very helpful. In my experience, this article is an extraordinarily accurate mirror of the spirit of the organizations that it treats and the attitudes of the scientists concerned.

These laboratories turn out to be very human organizations. In recruitment, for example, like prefers like, but in operation, the unique talents of various nationalities mesh in creating a whole that is greater than the sum of its parts. Using the evidence gathered in this study, a laboratory administrator could assemble teams of scientists from different nationalities to get optimum production, granting a certain equality of basic talents. Apart from its practical value, the study is fascinating in its own right, in withdrawing the veil from scientific research organizations to reveal the people who are engaged in these esoteric pursuits to be Germans, Frenchmen, Italians, or Britons, who also happen to be scientists.

The edifice of modern science was constructed by many different nationalities. Although Italy was its first home, it is impossible that it could have developed in Italy alone. The English, Germans, French, and Russians each had an almost unique contribution. The history of science has yet to be studied from this cultural point of view.

The international laboratories place the participating scientists in a new environment. The permanent values of this environment are still to be evaluated, yet one can already discern the effect on the younger scientists who move from their own countries to CERN and then to still another country, like the wandering scholar of an earlier time with his universal culture and universal language, Latin—today replaced by mathematics and science.

4 POLITICS AND INTERNATIONAL LABORATORIES: A STUDY OF SCIENTISTS' ATTITUDES[1]

Albert H. Teich

INTERNATIONAL SCIENCE AND INTERNATIONAL POLITICS

Functionalism and the Logic of International Laboratories

Changing relationships between scientific research, technological progress, and political power have produced radical tranformations in the scope and nature of international scientific cooperation. Where formerly most such efforts consisted of privately or academically sponsored exchanges of data, research results, or personnel, today one may observe the proliferation of large, intergovernmentally supported research establishments, bringing together hundreds of scientists from different nations and requiring substantial investments of capital and equipment. It is the needs of "big science" and "the technology of research" that have led to the establishment of such joint facilities. In such fields as space exploration and nuclear physics, the financial and manpower costs of doing frontier research are beyond the means of all nations except the superpowers. Hence, relatively affluent small and medium-sized nations with strong scientific traditions and close geographic and cultural proximity—such as the nations of Western Europe—are led to pool their resources in order to remain scientifically competitive.

It has long been a cherished belief among scientists that science is an ideal medium for international cooperation, and further, that such cooperation may lead to better understanding between nations and the reduction of world tensions.[2] One could cite an almost limitless number of scientists' statements containing the spirit typified by the well-known "Vienna Declaration of Scientists":

The ability of scientists all over the world to understand one another, and to work together, is an excellent instrument for bridging the gap between nations and for uniting them around common aims.[3]

On a European level, this belief meshes conveniently with the "European spirit," and cooperation in science and technology is commonly viewed as one of the instruments through which closer political integration of the various nation-states might be achieved. The line of thinking that is used to support this view is vague, yet highly seductive. It is seldom fully elaborated, but in its essence it corresponds roughly to the functional theory of political integration espoused by such scholars as Ernst Haas.[4]

There are four general means by which functional integration is supposed to operate, and opportunities for international scientific organizations can be seen in each case. First, a set of decisions is removed from the domain of international politics and

turned over to a specialized body with certain supranational powers. International
scientific ventures could conceivably make it possible to arrive at decisions involving
allocation of scientific resources and the distribution of effort across member states
in certain fields. Second, the lessons learned in a given sector may be applied to
other sectors. Thus, either by providing an example of successful operation or by
stimulation of its own requirements for, say, materials or legal authority, an international laboratory might encourage integrative efforts in other sectors. (This
is part of the widely held notion of "spillover.") Third, task orientation is favored
through the participation of technical experts rather than political actors. Within
large-scale international laboratories, scientists can work together on common
problems, unhindered by considerations of national background and political loyalty.
Fourth and finally, the existence of a supranational entity provides the basis for the
development of a larger set of loyalties that may be superimposed on national alle-
giances. This is of particular relevance to international laboratories: since it is
widely assumed that scientists are international in orientation, one outcome might
well be the evolution of a Europeanist technological elite of some influence, through
scientists' participation in such organizations.

Despite its potential in these four senses, it is probably fair to say that large-scale
international scientific cooperation in Europe has not yet had any marked effect on
European political integration. The most ambitious project (EURATOM) has been a
great disappointment, falling victim, practically since its inception, to one crisis
after another. The reasons for the short-run failure of these efforts have been analyzed with considerable insight by several authors.[5] Basically what they make clear
is that cooperative ventures in scientific/technological fields cannot have direct effects
on political integration without becoming politicized themselves. In the functionalist
sense, the beauty of employing scientific cooperation for the furtherance of political
integration hinges on the international, nonpolitical nature of science. This nonpolitical
condition can obtain only when the science is seen by national governments as being
essentially irrelevant to vital policy goals. Where it becomes relevant, a "logic of
diversity" (in Stanley Hoffman's terms) is likely to prevail.

There are, however, more subtle ways in which the political effects of such international scientific ventures may be felt. These derive primarily from the last of four
means of functional integration--the role of the scientists themselves. Over the longer
term, a number of sizable international laboratories operating stably within the European scientific community could have significant effects on the organization of that
community, its patterns of communication and mobility, and its overall strength and
vitality. This, in turn, could make that community a more potent political force, in

terms of fighting for its own interests, as well as its ideals. At the same time, the
growing link between scientific and technological power and the public functions of
national defense, economic growth, and satisfaction of social welfare needs is leading to a significant increase in the number and importance of scientific and technological advisors at all levels of government. Through its penetration of national
governments in such posts as well as in other political and administrative roles, a
scientific and technological elite of Europeanist orientation might well influence the
political process and foster political integration.

The Nature of this Study

Most of this is, of course, speculation. In order to deal in a more meaningful manner with the questions involved, it is necessary to look at the international laboratories with a special eye toward their political character. That is essentially what
this study attempts to do. The central question the study asks is "what is the nature
of the international laboratory environment?" In seeking answers, it follows two
basic lines of investigation. First, it looks at the laboratories in an operational
sense to see how they function as research environments. What special rewards do
scientists seek there and to what extent do they find them? How do national cultures
and the international nature of science interact on a day-to-day basis? What kinds of
operational generalizations can be drawn across the several different international
laboratories that exist, and how do these relate to the special situations of each?

Second, the study examines the political ecology of the laboratories. It looks primarily at the political attitudes of the scientists and engineers who inhabit the laboratories and asks several types of questions. What is the general framework of
political orientation among the scientists? To what extent is internationalism an
apparent trait? What are the distributions of opinion on key political issues, and how
are these distributions related to characteristics of the scientists? How do these views
compare with the expressed views of other scientists and other elite groups? What
is the place of politics in the overall laboratory picture? And, finally, how are political attitudes, particularly those relevant to European integration, affected by the
international laboratory experience?

Our means for seeking the answers to these questions draws upon an attitude survey of some 320 scientists and engineers from Europe's three largest and most important international laboratories: ESTEC (the European Space Technology Center),
CERN (the European Organization for Nuclear Research), and ISPRA (the EURATOM
Joint Research Center at Ispra). Both written questionnaires (in English, French,
and German, about two-thirds of the sample) and personal interviews (in English and

French, about one-third of the sample) were employed.[6] While the sample was not randomly drawn, its basic outlines on the critical parameters of laboratory affiliation, nationality, and length of service (seniority) roughly parallel the structure of the population from which it was drawn, and its size is large relative to the size of that population (approximately 20%). On this basis, extrapolations to the class of international scientists and engineers do not seem unreasonable. Table 1 presents the overall distribution of the sample by its major parameters.[7]

EUROPE'S INTERNATIONAL LABORATORIES

Although CERN, ESTEC, and ISPRA share the important aspect of being international in sponsorship, their fields of work, aims, accomplishments, historical contexts, physical appearances, and atmospheres vary enormously. To consider these laboratories as completely unrelated would be to miss the essential point that in being international they share a peculiar quality which sets them apart from other scientific establishments—a quality of which the staffs are highly conscious. Yet, in failing to take note of the differences between these centers, one loses an important dimension along which the attitudinal data must be explored. This section, therefore, outlines briefly the natures of the three laboratories, their histories, and some of the basic data needed for understanding them as they were at the time of the study.

High Energy Physics: CERN

CERN, the European Organization for Nuclear Research, is the oldest and most prestigious of the major international laboratories.[8] It is generally recognized as the major triumph of European scientific cooperation and it is the standard to which all of the other international laboratories, fairly or unfairly, are compared. The exclusive focus of CERN's efforts is high energy physics: basic research on the nature of matter, through studies of subnuclear particles. The laboratory was founded in 1954 and is located just outside of Geneva, Switzerland. In 1967, it employed a total staff of approximately 2,350 persons (while hosting some 400 nonstaff vistors) and operated on a budget of about $40 million. The organization is sponsored by 12 member-states which contribute to its support in proportion to their net national incomes.[9]

The initiatives that led to the founding of CERN resulted from private discussions among several influential members of the international physics community, beginning around 1948.[10] The scientists were concerned about the ability of the European nations to carry on research at the forefront of nuclear physics because of the need for new, large, and extremely expensive particle accelerators that no single European nation could reasonably afford to build and operate alone. At the same time, many

Table 1. Respondents by Laboratory, Nationality, and Seniority

	ESTEC					
	Br[a]	Fr	Ge	It	Other[b]	Total
Low Seniority[c]	24	14	15	4	15	72
High Seniority[d]	11	5	5	3	11	35
TOTAL	35	19	20	7	26	107
	CERN					
Low Seniority	11	7	10	10	7	45
High Seniority	14	6	7	9	16	52
TOTAL	25	13	17	19	23	97
	ISPRA					
Low Seniority	0	2	9	8	1	20
High Seniority	2	23	26	31	14	96
TOTAL	2	25	35	39	15	116
	OTHER[e]					
Low Seniority	2	3	9	0	15	29
High Seniority	5	1	8	6	15	35
TOTAL	7	4	17	6	30	64
GRAND TOTAL	69	61	89	71	94	384

[a]Abbreviations Br, Fr, Ge, It are used in tables throughout to represent Britain, France, Germany, and Italy.

[b]Includes: Holland (24), Belgium (21), Switzerland (13), Austria (12), Sweden (4), Denmark (3), Norway (8), Finland (2), and Spain (7).

[c]Less than 2 1/2 years in lab (166).

[d]More than 2 1/2 years in lab (218).

[e]Includes IAEA (11), ESDAC (18), PETTEN (10), HALDEN (20), and FONTENAY (5). (See footnote 6.)

European political and intellectual leaders were actively interested in developing
tangible demonstrations of European cooperation. Under the institutional sponsorship of UNESCO, the confluence of these two drives produced an agreement establishing a provisional organization two years later. In fact, it is typical of CERN
that, by the time the permanent organization legally came into existence, the laboratory was already a going concern with a staff of 120.

In the years since 1954, CERN has emerged as one of the world's leading institutions in high energy physics. At the time of this study, the capabilities of its 28 GeV
proton synchrotron were matched only by the machines at Brookhaven National Laboratory in the United States and the Dubna Joint Institute for Nuclear Research in the
Soviet Union. CERN has a strong internal staff, but its program also stresses participation of scientists from outside institutions. Experiments may be carried out by
CERN staff teams, teams of visitors from European universities or institutes, or
mixed CERN-visitor teams. In addition, CERN provides a great deal of raw data,
usually in the form of bubble chamber photographs, to various research centers in its
member countries.

CERN is blessed with an extremely beautiful location, just outside of Geneva, in
the shadow of both the Alps and the Jura Mountains. Its site provides simultaneously
the pleasanter aspects of pastoral life: quiet, clean air, and a generally relaxed informal feeling, and (within a short drive) much of the variety and excitement of an
international city. Few of CERN's scientists are dissatisfied with life in the Geneva
area.

Morale at CERN has always been high, and few employees have anything but praise
for the laboratory and, in general, the way it is run.[11] Alone among the three laboratories studied, CERN operates on a 24-hour-a-day basis. The product of economic
necessity, since anything less would be terribly wasteful of the vast investment represented by the accelerators, this schedule adds to the informality of the CERN scene.
Scientists, especially those with an experiment in progress, are almost as likely to
be found in their laboratories at midnight as they are in the afternoon.

One of the most important aspects of CERN's intellectual atmosphere is the presence
of numerous "name" physicists at the laboratory. Virtually everyone who is anyone in
European high energy physics is in some way associated with CERN. The presence of
men with such "intellectual sex appeal" is a great attractant for younger scientists,
and many come not only to use the machines but to work under and learn from these
men. In this and other ways, CERN clearly gives the impression of being in the
scientific "mainstream."

Nuclear Power: EURATOM-ISPRA

The ISPRA laboratory is the largest of the four Joint Research Centers of EURATOM, the European Atomic Energy Community. In many respects, it is almost the complete antithesis of CERN. Organized primarily for political purposes and tied closely to sensitive economic and military interests, EURATOM has been beset since its birth with a host of difficulties. The bureaucratic wrangling and continuous budget crisis affecting EURATOM have prevented ISPRA from achieving any meaningful scientific or technological successes and have reduced its morale to a pitiable state.

Together with the ECSC (European Coal and Steel Community) and the EEC (European Economic Community—the Common Market), EURATOM is a part of the European Community (EC). (At the time of the study the EC consisted of the so-called "Six": France, Germany, Italy, Belgium, the Netherlands, and Luxembourg. Now, of course, it has expanded to include Great Britain and several other countries.) Its origin dates from the failure of the plan for a European Defense Community in 1954 and from subsequent efforts of the "Europeanists" to maintain momentum toward their long-range goal of political integration. The rationale for EURATOM, which was to have been a vehicle for pooling the nuclear power development efforts of the Six, was based on a number of economic planning studies that forecast rapid growth in Europe's energy needs and a shortage of fossil fuels.[12] In reality, the need for nuclear power has never lived up to these forecasts, and EURATOM, rather than serving to pool the efforts of its members, has been superimposed upon them as a competitor for manpower and financial support.[13]

In any case, EURATOM was established by a treaty concluded in 1957, which empowered it (in part) to control nuclear materials; encourage business enterprise in its member states; facilitate free movement of nuclear materials, capital, and personnel; place research contracts; and establish its own "Joint Nuclear Research Centre" [which, in fact, consists of four separate establishments: ISPRA (Italy), PETTEN (Holland), KARLSRUHE (Germany) and GEEL (Belgium)].

EURATOM's research program became operational in late 1960, and squabbling among its member states, over budgets and other matters, began almost immediately. At the time of this study, in mid-1967, EURATOM was in the throes of a major budget crisis, with the council totally unable to agree on a third five-year program. Although agreement was finally reached, the organization has never really recovered from this crisis.

As the largest of EURATOM's in-house laboratories, ISPRA has been most sensitive to EURATOM's difficulties. The center was established in 1957, as part of the Italian national nuclear program, and transferred to EURATOM in 1961. By mid-1967, it employed some 1,600 persons and had an annual budget estimated at $17 million per year.[14] Beyond direct reactor development work, the laboratory's program includes research in neutron physics, metallurgy and ceramics, data processing, and radio-biology. The laboratory is located on a wooded, 400-acre tract, just outside the tiny village of Ispra on Lago Maggiore, some 45 miles northwest of Milan.

EURATOM's difficulties, combined with some special characteristics of the ISPRA situation, have created a very unusual atomosphere in the laboratory. By 1967, the staff had been subjected to the organization's traumas for so long that it no longer cared enough to be militant. Rather, the scientists were quietly and sadly aware that matters outside of their control would determine the fate of their organization and their work. They held some mild hope for the future but not much enthusiasm. Most of all, they felt that their work and the real scientific capabilities of their center were going unrecognized because of irrational political disputes.

Several other dimensions of ISPRA's character are worth noting. First of all, despite its transfer from the Italian government to EURATOM in 1961, ISPRA in 1967 was still very much an Italian establishment at heart. Nearly half of the laboratory staff, excluding locally recruited labor, were Italians. Beyond this, there is the fact that the center is placed in a highly parochial rural environment. If a foreigner at ISPRA wishes to have any contacts outside of the EURATOM community, he has no choice but to adapt to Italian ways. Although officially, ISPRA (like the entire European Community) has four languages, French, German, Italian, and Dutch, most of the center's business (formal and informal) is carried out in Italian. (Probably second in currency is English, a language that while not officially recognized by the EC, is widely spoken, especially among Germans and Dutch.)

ISPRA's setting is even more idyllic than that of CERN. From the grounds of the laboratory, one can see the snowcapped Alps rising majestically behind the azure waters of Lago Maggiore. The visitor to ISPRA cannot fail but to be impressed, even distracted, by the beauty of the surroundings, and, as at CERN, he must feel that even employees of long standing have not become oblivious to it. The village of Ispra is not Geneva, however. It is a hamlet of some few hundred residents, and since the only large city in the region, Milan, is 45 miles away, the staff lives primarily in a rather rural area, where the variety and diversions of an urban environment and the prerequisites for a cosmopolitan life-style are almost totally lacking.

In attempting to compensate for these defects of the ISPRA locale, EURATOM has created a self-sufficient community that is largely independent of the indigenous population. There are some permanent apartments on the site, as well as a clubhouse that serves as a social center for after-hours activity. There is a European Community school in a nearby town, a EURATOM beach at the lake, and a EURATOM general store on the site. Concerts, movies, clubs, and lectures are available for those who are interested. ISPRA is perhaps the only international laboratory (or national laboratory for that matter) that has its own miniature golf course.

Despite this tightly organized environment, ISPRA has had some rather severe morale problems. These stem in part from the difficulties of the community, the fact that funds have been in short supply, the fact that no one is really sure what the ultimate aim of all this reactor research is, and the fact that the future course of the organization is completely unknown. Uncertainty about the future is compounded by the fact that most of the EURATOM staff have what the Europeans call "fonctionnaire" status. This is roughly equivalent to a civil service appointment in the United States, and implies a high degree of job tenure. Naturally, a secure job in an organization whose very existence is in doubt is not conducive to high morale.

Some of the consequences of this morale problem have been unique in the realm of scientific organizations and are worthy of a study in themselves. For example, most large laboratories have some sort of gate to restrict access to their sites. For employees, passage through these gates is nothing more than a formality. At ISPRA in 1967, however, the gates were kept locked day and night except for a short period in the morning, lunch time, and a similar short period in the evening. All employees (regardless of position) who arrived more than 15 minutes after the starting time in the morning were not allowed to enter without giving their names to the guard and filling out a "tardy slip," which was eventually forwarded to their supervisor. In the evening, no one was allowed to leave until the official quitting time, and the long line of automobiles in a queue behind the gate, with their engines running ten minutes before quitting time, was a most amusing sight to the visitor. The decision to keep the gates locked, it is said, was occasioned by the fact that employees were taking advantage of the normal professional freedom. According to one administrator, there were some who "were only showing up to collect their checks."

Only in a few isolated pockets of the center did one find relatively high morale, those areas that were more peripheral to the center's main foci and therefore less dependent upon its management. Except for these areas, the type of intellectual excitement that was apparent at CERN did not exist at ISPRA. Despite some recruitment difficulties, ISPRA managed to accumulate some excellent people. But organizational problems prevented the center from becoming in any sense a place where bright young scientists and engineers might wish to come and learn from top people in their fields.

ISPRA's physical isolation, the hostility of "competing" national programs, the fact
that there is very little turnover among the personnel, and the fact that ISPRA at-
tracts few short-term visitors, combined, finally, to give ISPRA people the impression that they were rather isolated and outside the mainstream of activity in
their fields. [15]

Space Technology: ESTEC

ESTEC (the European Space Technology Centre) in Noordwijk, Holland, is the new-
est of the major European international laboratories. The site of most of the in-
house research and engineering carried on by its parent organization, ESRO (the
European Space Research Organization), ESTEC came into existence in early 1963
but by mid-1967 was not yet fully operational. ESTEC has suffered from the political
and administrative infighting that has plagued ESRO, but its problems are nowhere
nearly as serious as those of ISPRA.

ESRO had its origins in a series of informal discussions among a small number
of influential Western European scientists in the late 1950s. Concerned about the
European role in space and encouraged by the success of CERN, this group (several
of whose members had been involved in the early stages of CERN) developed the basic
concept of a joint European space effort. In many ways, these early discussions were
shaped by the personalities involved. One insider recalls that many of the key ideas
emerged from conversations during a pleasant, and slightly tipsy, evening stroll
through the streets of London. Another characterizes this stage (not at all in a dis-
paraging way) as a nostalgic attempt on the part of several scientists to recapture
the spirit that prevailed when European scientific collaboration—and they themselves
—were a decade younger.

In any event, the ESRO idea entered the public domain when, at a COSPAR[16] conference in January 1960, Professor Pierre Auger of France became its "promoter."
Within a few months the governments of several Western European nations had become sufficiently interested to warrant the holding of more formal meetings and the
establishment of a study committee. It took some time for the nations involved to
reach agreement on the many complex issues raised, and nearly three and one-half
years elapsed between the first formal meeting of delegates and the coming into force
of the ESRO Convention in March 1964. [17]

The purpose of ESRO, according to Article II of its Convention, is "to provide for
and to promote collaboration among European States in space research and technology,
exlusively for peaceful purposes."[18] To this end, the Convention provides that ESRO

may design and construct sounding rocket payloads, satellites, and space probes; procure launching vehicles and arrange for their launching; support needed research and development; and perform various other specified functions. The organization maintains several smaller facilities in addition to ESTEC and, following the NASA model, contracts out much of its work. ESRO's original budget provided for the expenditure of about $300 million over an 8-year period. Its 1967 budget was $48 million. As a scientific body, ESRO has operated mainly in the form of a service organization, catering to the experimental needs of national groups. These groups, which may come from universities or government laboratories in member countries, propose experiments for ESRO to carry on its space vehicles. ESRO then builds the hardware (usually on contract), integrates the payload (at ESTEC), arranges the launchings (to date, through NASA), operates and monitors the vehicle, and supplies the data to the group that requested it. Although at the time of our first visit to ESTEC (April 1967), the organization had not yet attempted any satellite launchings, since then it has had several significant successes. It has also carried out numerous sounding rocket experiments.

For a number of reasons, one could best describe the condition of ESTEC in April-May 1967 as that of a state of flux. The newness of the top management (the director, head of administration, and head of personnel had all been replaced within the past several months), the release of a major report on ESRO reorganization, and the physical moving of the laboratory from temporary quarters in Delft to its permanent site in Noordwijk, all contributed to this impression.

Although ESTEC was not officially established until the ESRO came into being in early 1964, a staff nucleus had been working in Delft since January 1963. Despite this head start, the establishment was rather slow in reaching full operation. Construction of the permanent buildings proceeded less rapidly than planned, and a fire in some temporary buildings set things back even further.

At the time of the study, about half of the establishment was located in Delft (in the Technical University), some offices were located in a hotel in Noordwijk, and the remainder of the establishment had already moved into the permanent buildings. At this time, ESTEC employed a staff of 464, of which 321 were considered to be of professional grade.[19]

To the extent that a single nationality can be said to give a "flavor" to an establishment, ESTEC, in mid-1967, had a distinctly British flavor. About one-fourth of the entire staff complement at ESTEC was British, and the English language was used almost exclusively in conversation.

One aspect of the ESTEC atmosphere which struck even the most casual observer was the general lack of intellectual excitement in the air. To a visitor accustomed to a more or less academic atmosphere, where a deep commitment to one's work is the norm, this was especially striking. Perhaps the comparison is unfair, since ESTEC is avowedly "technological" in nature rather than "scientific" and its work tends to be engineering and development (hardware-oriented) rather than research (knowledge-oriented). In any case, the atmosphere did not seem to be that of a creative establishment. There were no name people at ESTEC—"technological stars" one might call them—who could attract bright young scientists and engineers. There was little after-hours work, and there appeared to be little internal professional activity at ESTEC which was not directly related to one's job.

There seemed to be a general feeling that the choice of location was a mistake. Holland is not seen by most other Europeans as an especially interesting country or one with a very pleasant climate. Noordwijk, in particular, is a summer resort town on the North Sea coast, with a population of about 20,000, with virtually no cultural or intellectual life. The site on which ESTEC is located abuts a row of sand dunes, on the other side of which is the sea. Blowing sand and strong winds were constant sources of irritation.[20]

Overall, however, there was a great deal of hopefulness evident in the ESTEC atmosphere in spring 1967. Everyone spoke of the mistakes of the past several years, but most were willing to attribute the bulk of these mistakes to ESTEC's "running-in" period. Several persons who had held responsible executive posts were viewed as catastrophic failures—whether this was due to their own ineptness or constraints placed upon their offices by the organization's structure is an open question—but their replacements were looked upon more favorably. The impending completion of ESTEC's permanent building was seen as an important step toward "getting settled." The launching of the first satellite, giving the world the first tangible evidence of ESRO's accomplishment, was anticipated as a morale booster.[21] Finally, the Bannier Report (on reorganization) was seen by most as giving official sanction to general gripes that "we had known about all along" and therefore providing commonsense solutions to many of the organization's troubles.

NATIONAL STYLE AND INTERNATIONAL SCIENCE

Who Are the International Scientists?
Despite these profound differences among the three laboratories, in several ways, the respondents in this study were a rather homogeneous group: more than 95% of them

were male. Their average age was 35 and, in fact, nearly 60% were in their thirties. Most held an educational degree equivalent to something between an American bachelor's and master's degree. Only 6% held the equivalent of an American Ph.D., while another 8% fell between the M.S. and Ph.D. equivalent levels. In terms of grade within their organizations, most were at middle levels. More than half were either physicists or electrical engineers, but the fields of mechanical engineering, chemistry, metallurgy, mathematics, computer science, and biology were also represented.

Digging a bit deeper, however, some patterns appear that suggest differences in the types of scientists and engineers drawn to the various centers. For example, at ESTEC, where the laboratory as well as the sample was weighted with a large proportion of Britons, there were more engineers, few academic types, and many individuals recruited directly from government and industry. ESTEC was also the only one of the large centers where a significant percentage (15%) of the respondents had spent more than ten years in their previous jobs. CERN, on the other hand, was the most academic in its orientation, with the largest percentage of respondents coming directly from the government and industrial sectors.

A number of other interesting differences emerged in examining the processes through which the scientists joined the international laboratory. It appears that CERN respondents were not particularly dissatisfied at their previous jobs but chose to leave because of the "better opportunity" or "broader experience" CERN offered. A plurality of ISPRA respondents, on the other hand, left their former jobs because of career or intellectual dissatisfaction (20% and 21%, respectively). At ESTEC, finally, nearly one-third of the respondents (30%) left their previous jobs because of dissatisfaction with career progress.

With regard to reasons for choosing the present organization, responses are similarly revealing. Table 2 shows the relative importance of the various motives reported. The strength of professional orientation at CERN is evident here. CERN's staff had come to Geneva to do high energy physics at the largest accelerator in Europe, or among those who were not high energy physicists, at least to follow a line of work which interested them. For most CERN respondents, other considerations were secondary. Of particular note is the fact that CERN's international makeup, while (as will be seen later) valued by nearly everyone, was not rated as an important reason for coming to CERN in the first place. In contrast, ESTEC's personnel were attracted not only by the nature of its work but also by the higher salaries it offers (particularly relative to those in Britain) and by the fact that it is an international center. One might conclude that such initial expectations in the minds of ESTEC staff

Table 2. Relative Importance of Reasons for Coming to the Organization, by Laboratory[a]

	ESTEC	CERN	ISPRA
Opportunity to Pursue Particular Type of Work	high	high	high
Desire to Work with Particular Individual or Group	low	low	low
Quality of Equipment	low	medium	low
Higher Salary	high	low	medium
Desire to Work with People of Other Nationalities	high	low	medium
Location	low	low	low
Job Security	low	low	low
Lack of Opportunities in Own Country	low	low	low

[a]On this question, respondents were asked to rate each possible reason as "very important," "a factor," and "not important." In addition, they were asked to select one reason which they felt was "most important." This tabulation was prepared by considering a reason "high" in importance for a laboratory if 40% or more of the respondents rated it as "most important" or "very important." A reason was ranked "low" if 40% specified it as "not important" or did not rate it at all. Reasons that fitted neither the high nor the low criterion were considered "medium."

members imply a low degree of professionalism and greater degree of commitment to nontechnical (career and sociopolitical) goals.

Working Together

Most of the scientists and engineers who come to the international laboratories do not come entirely unprepared for the cross-cultural experience. The data reveal, in fact, that they are a rather cosmopolitan group. Nearly all reported having traveled substantially in Western Europe, and 40% had been to the United States or Canada. Even more important, nearly half (48%) reported having lived in a foreign country (for three months or longer) prior to coming to their present jobs. The scientists also showed a remarkable facility for languages—they spoke an average of 3.3 languages. Only 17 persons in the entire sample (4%) were unilingual (all 17 were British, 14 from ESTEC). Fifty-two respondents spoke five or more languages.[22]

These factors facilitating cross-cultural communication ought to bear considerable importance, since, as scientific working environments, the international laboratories

are unique in Europe, and in some respects, the world. When scientists go to work
in a conventional laboratory in a foreign country they tend to adapt themselves to the
vague set of operational habits, customs, and values that constitute the national style
of life in their host country. In the United States, for example, they might be expected to speak continually in English, eat sandwiches for lunch, deal informally with
their superiors, and so forth. While some alterations in the behavior patterns of the
resident Americans might be observed (for example, speaking more slowly and
distinctly in order to be understood) it is clearly the foreigners who do most of the
adapting. At ISPRA, CERN, and ESTEC there are no real "natives" whose behavior
represents the norm and acts as a basic pattern for others. The need for adaptation
is common to all, and a central meeting ground must be found that, while drawing
on the characteristics and styles of its various components, is distinct from them
all.

Finding this central meeting ground is, of course, greatly facilitated by the subject matter of science. Common research problems and common methods of work
are the forces that tend to place the activities of scientists in a context preceived
as largely independent of cultural and societal differences. It is naïve to think, though,
that nationality plays no role in the research process or that it can be ignored as a
factor in the operation of these laboratories. To quote one respondent,

The research, the methods of research are quite the same. There is no national
physics; the physics I studied in Germany, I continued to study in the U.S.A., and
I practice here [in ISPRA] are all quite the same. The thing that is different is the
way different people attack a problem....I mean, a solution is a solution, but how
you get to the solution, this depends on your background, your cultural and national
background.

The fact of being born and raised a "Briton" rather than a "German" implies a whole
set of cultural and historical assumptions that are constantly manifest in behavior
patterns. The operation of an international laboratory requires the continuous jux-
taposition of these patterns.

A fundamental lack of prejudice is essential to the scientific ethos. The belief that
nationality, race, and religion are irrelevant, and the quality of his work is the only
basis on which to judge a fellow scientist pervades and helps to maintain the international scientific community. The functionality of this belief has been demonstrated by
the experiences of science in those societies, such as Nazi Germany, which rejected
it.[23] Sometimes, however, a scientist's strict adherence to this unbiased outlook leads
him to conclude that a scientist's nationality does not affect his style of work. An international laboratory provides an environment where such an illusion may not long
survive.

The interviews revealed that virtually everyone observed national differences in
the styles of work of his colleagues. Most often the scientists reported that the widely
recognized national stereotypes, although rather too general, had some validity after
all. For example, to quote one CERN Frenchman,

I mean you might sometimes find a well-organized Italian and a very fast Dutchman,
but the reciprocal is more often true....I mean all these categories are a little absurd, there might be exceptions, but these so-called national characters exist.

As did this respondent, most of the subjects cited the contrast between the Latin and
Anglo-Saxon/Germanic temperaments. Their descriptions were informative and
sometimes amusing, and it is worth citing a number of them here in order to catch
the flavor of expression. A Swiss engineer who was concerned with staff recruitment
noted the classic Latin-Nordic contrast and spoke of its relation to employment interviewing:

There is another difficulty and I'd like to bring it up now. This is the difference in
character. The two extremes could be the Swedish people and the Italian people. The
Swede is very calm, very discreet, very slow in speech, whereas the Italian is exactly the opposite. So when we have interview boards and when we go from one person to the next, it really takes an effort to realize we are talking to a totally different
person...and try to assess both people correctly.

Generally, the traits associated with different nationalities were seen as becoming
part of the national scientific traditions. Latins were credited with being more theoretically oriented, abstract, mathematical, and imaginative. Northerners, on the
other hand, were considered more practical, persevering, physically oriented,
and thorough. [24] A high-ranking German physicist from EURATOM gave a graphic
description contrasting his own countrymen with the French:

There is a conflict particularly between the Germans and the French. The Germans
are trained to take a very experimental approach, a physical approach to physics,
while the French always look at things mathematically. Do you understand me? The
Germans make physics from a model. Angular momentum, for example, of an elec-
tron, of a particle, is seen as spinning. [Interviewee made hand motions to demonstrate.] It's not just a mathematical symbol or an equation.

On the other hand, a French mathematician provided perhaps the simplest way of
distinguishing between European mentalities:

Of course, the differences in mentality exist.... For example, you can really draw
a line cutting Europe in two parts: the part where the shops are closed after six, and
the part where the shops stay open after six. In the first one you have Great Britain,
Holland, Germany, Switzerland, and all the northern countries. In the other part you
have Belgium, France, Italy, Spain, and so on.

Although we lack quantitative data to support them, on the basis of the interviews we may propose two rather general assertions with regard to these stereotypes. First, while it seems that the images of other nationalities are nearly always given with positive or neutral affect, there is little tendency for them to disappear with increasing experience at the centers. In fact, there is some evidence to suggest that those individuals with little prior foreign contact who had at first rejected stereotypes as "unscientific" seemed to discover, with experience, that they really do exist. Second, the saliency of nationality, and hence of these images, tends to decrease with the professionalism of the scientist.[25] This is not to say that highly professional individuals did not maintain national images; some of the most colorful and articulate descriptions came from the mouths of eminent scientific personalities. Rather, with increasing professionalism, the scientists tended to perceive the special field, school, or place of previous work as stereotyping features of equal or greater importance than nationality. For example, a colleague might be viewed as a former student or co-worker of Professor Amaldi at Frascati (which implies a set of special professional attributes) rather than as a "demonstrative" or "emotional" Italian.

Morale

One of the striking features common to all of the laboratories included in this study was the virtually total absence of any staff conflicts along national lines. It was, of course, a source of pride to many of the scientists to whom we spoke that tendencies to coalesce on the basis of nationality did not appear. Conficts concerning, for example, the allocation of machine time at CERN, rules and regulations at ESTEC, or definition of the future program at ISPRA were far from absent in any of the laboratories. But they pitted one division or team against another or the scientific staff against the administration rather than one national group against another. This is not really too surprising in light of the fact that interests in such disputes are likely to be distributed according to position and not nationality. Further, simply because friction between national groups was an anticipated, and potentially inflammable, problem area, people took pains to assure that it did not become involved where it potententially might. In EURATOM, where conflicts between the policies of member governments toward the organization have been practically continuous—often finding France opposed to the other members—the personnel, whose own self-interest regularly differs from their governments' policies, have not reflected these disputes.

Several survey questions attempted to explore the similarities and contrasts between the international research environment and its national counterparts. The respondents were asked: "Are there any characteristics of this laboratory which might make it easier to perform research or other technical activities than in a national or university laboratory?" and "Are there any characteristics which might make it more difficult?"

It is clear that the responses are largely a function of the special state of affairs within each center and reflect, in each case, the atmosphere described earlier. Table 3 presents the data for these two questions. The contrast between CERN on one side and ISPRA and ESTEC on the other is evident, and the interviewees also expressed in their own words what the tabulations suggest. The answers to the "Easier and More Difficult" questions at CERN were almost stereotyped:

Well, CERN is unique for a start. It's the only place in Europe where you can do this high energy physics work.
—British physicist

First of all, CERN is a much bigger organization. It has got a budget of some 170 million Swiss Francs, while my institute in Italy had perhaps half a million. This is a hell of a difference!
—Italian physicist

The feeling, at least among the physicists, was one of expansiveness, of freedom. An administration sensitive to the needs of the scientists, willing to gamble on "long-shot" experiments, and lacking the usual bureaucratic traditions was often pointed to with pride.

ISPRA presents a rather different picture. Bureaucratic and political complications render the performance of research or other scientific activity a much more trying process than the respondents would like. [26] One interview excerpt typifies the resulting frustrations:

We are complaining here of the lack of management. I think it's true probably because now the European Community has come to a difficult point and there is no political will to continue, to go deeper. This lack of political strength at the top level of the Communities reflects down to this lack of management. There is no well-defined purpose to go to. ... You may be working with some colleagues who share your interest but, well, you do not feel you are part of some project which is building up.
—Italian physicist

Perhaps the most significant part of the ISPRA situation is the feeling that what the scientists actually do does not matter very much, that scientific performance is not the real criterion by which they will be judged. What does matter, community politics, is totally beyond their control. Being on the one hand scientists committed to their work and on the other hand "Europeans" committed to European integration, they cannot avoid a sense of impotence. In the revealing words of one ISPRA Frenchman,

Table 3. Ease and Difficulty of Performing Research, by Laboratory[a] (%)

	ESTEC (n=107)	CERN (n=97)	ISPRA (n=116)
Easier:			
Yes	44	73	48
No	43	12	44
Don't Know	13	15	8
More Difficult:			
Yes	64	37	65
No	21	45	25
Don't Know	15	18	10

[a]Percentages in this and subsequent tables are rounded, and totals may occasionally vary slightly from 100%.

This matter of [EURATOM] politics is not made at our level at all. In fact it is made at such a level that nobody can do anything about it.

ESTEC, finally, shows neither the expansiveness of CERN nor the frustration of ISPRA. The respondents here see problems but they are not nearly as serious as those at ISPRA. Mainly, they stem from organizational politics:

This is an organization that in principle is supposed to return some of the money invested by the members in the form of contracts. The effects of this, at least for the moment, are quite bad because it makes, at high levels, political conflicts that are sometimes far from technological and scientific interests.
—Spanish scientist

Bureaucracy is also a part of the ESTEC scene and the reaction to it is similar to that at ISPRA:

[The main difficulty] is certainly the fantastic mess you have from the administrative point of view. This is an algebraic addition of all the defects of national administrations and it is really something to see.
—French scientist

One result of the interview analysis is to suggest that, although the percentages of respondents who found advantages and disadvantages to the international environment were quite similar at ESTEC and ISPRA, the real feelings were rather different. Individual response to defects in the organization was much more severe at ISPRA than at ESTEC.

Over and above the special characteristics of each center, there is a major set of attitudes widely shared throughout the sample: reactions to the internationalism of the environment. Virtually all of the respondents remarked on the personal and professional stimulation of the international environment. In describing the personal

value of the stimulation that they received from the international environment, the
respondents often credited it with the same sort of vague benefits people claim to
receive from travel:

I think coming here and working with other nationalities has been quite a good experience; it broadens one's experience. ...You can't put your finger on it exactly
and say "now this is very good," but I feel it broadens one's experience.
—British engineer at CERN

If we had the same CERN in Italy staffed wholly with Italians I'd perhaps do the same
physics or the same mathematics and I'd enjoy these things professionally. But I
would miss this contact with other people who have other kinds of education, other
ideas in their heads, other prejudices than mine.
—Italian physicist at CERN

The reward is in part that of experiencing directly that which one has been accustomed to learning only from secondary sources. Further, in the international experience there is the simple joy of discovery—finding known facts or ideas placed
in new contexts or viewed in different ways.

Finally, one part of the reward of the international experience appears to be that
of making one's own self-image more cosmopolitan. Within the social system of
science, cosmopolitanism is a highly valued trait. By comparison with their reference groups at home, these respondents feel themselves far more experienced and
sophisticated. A French physicist from ISPRA, who obviously left himself open to
being called a snob, put it this way,

I believe that my fellow countrymen who have stayed in their own country for all of
their lives cannot have a very objective way of judging many problems. They are
biased and this is unconscious, I would say. It is a very amusing experience when you
go back to your own country and discuss some problems to see how people are biased.
They never lived outside their own country and were never exposed to different points
of view.

The feeling that one has been "broadened" seems to mean a good deal to the respondents and the way in which they view themselves. It is common to the several nationalities in all of the laboratories included in the study.

Several dimensions of the international work experience, are incorporated in the
question of professional satisfaction. Table 4 shows the distribution of responses to
the question, "How satisfied would you say you are with the way things have developed
for you [in this organization]?"

In light of the scientists' reactions to the atmosphere of each center, there are no
great surprises to be found in the levels of satisfaction reported. The CERN sample
shows the highest level of job satisfaction. Both ISPRA and ESTEC show smaller
percentages at the upper part of the scale—only one respondent from either laboratory

Table 4. Job Satisfaction, by Laboratory and Nationality (%)

	ESTEC				CERN				ISPRA		
n =	Br (35)	Fr (19)	Ge (20)	It (7)	Br (25)	Fr (13)	Ge (17)	It (19)	Fr (25)	Ge (35)	It (39)
Completely Satisfied	0	0	5	0	16	8	12	5	0	0	0
Very Satisfied	40	21	5	14	48	46	59	37	32	20	18
Satisfied	23	53	50	14	20	38	18	47	20	46	44
Somewhat Dissatisfied	34	21	35	43	12	8	12	11	36	26	28
Very Dissatisfied	3	0	5	0	4	0	0	0	12	6	8

said he was completely satisfied with his position—while substantial numbers appear at the lower part of the scale. ISPRA, in particular, shows a significant few at the extreme of dissatisfaction.

Career Patterning

It appears that one factor which determines to a large extent the degree of satisfaction which an individual receives from his stay in an international laboratory and which consequently, in aggregate, strongly affects the morale of the laboratory is the degree to which the scientist sees his stay as permanent. This is a rather complex matter. In essence, it seems that the more strongly and more long-term one commits himself to one of these organizations, the more sensitive he is to defects in the environment and to organizational instabilities. This is perhaps more a factor of the scientist's state of mind than of the real length of his stay. Many of the scientists at ISPRA have been there relatively short periods—on the order of five years —while many of the CERN staff members have held their postions at least twice as long. At ISPRA, however, the scientists are making semipermanent decisions in coming to the organization; they generally receive tenure after about two years and many, when they began, expected to be able to remain in the organization effectively for the duration of their careers. At CERN the organization's intent is to give as many qualified scientists as possible an opportunity to be a part of the organization. Hence, an individual generally starts out with a three-year contract. This may be renewed for three more years, and after the renewal, if the individual wants to stay and the organization still wants him, an indefinite contract may be given. The difference is in the incremental nature of this commitment. One does not see his choices as long-term, rather they are made stepwise. One consequence, reported by many CERN respondents (even those who have indefinite contracts), is that they never feel quite permanent in Geneva. In the words of one British scientist:

I don't think we ever really feel permanent here, that we are going to stay permanently.... One has the feeling, okay, one's been here five years, perhaps I'll stay a few more years, I don't know, perhaps not. But it's still the feeling, this isn't our permanent home.

Although the feeling vaguely disturbs many of the CERN respondents—the above-quoted individual mentioned later that he believed some of the value of the experience is lost because of the sense of transcience—it seems to be largely functional, in that it allows the scientists to put up with many minor problems, both in and out of the organization, which might be considered more serious if they were seen in a permanent framework.[27]

While CERN and ISPRA represent relatively pure cases, ESTEC is too new to fit either pattern. Contracts there are of limited duration and many of the staff members (particularly the British) are on leave from government establishments. The net result is certainly far from a permanent commitment on the part of the staff, and in fact the other extreme—a commitment too weak and temporary in nature—may have been approached. We were told that some respondents, Frenchman and Belgians, leave their families at home and commute back and forth on weekends. This situation of transcience—leading one French respondent to reply, when asked about life in Noordwijk, "I must say quite frankly, I don't live there, I just work there"—appears to yield a level of commitment to the organization which is less than optimal.

THE PLACE OF POLITICS

The Depth of Political Involvement

It is apparent that the vast majority of scientists and engineers at the international laboratories maintain a definite interest in the political world.[28] In some cases, however, the particular word "politics" provoked resistance in the subjects. This was true especially at CERN, perhaps the most avowedly apolitical of the organizations, but it also happened elsewhere. Experience showed that "international developments," "public affairs," "current public issues," or any other such terms that describe the larger political domain, seemed to be more generally accepted. The explanation for this phenomenon appears to be a semantic one, in that the word "politics" alone carries, for some scientists, connotations of local party activities, campaigning, banner-waving, and personal bargaining—activities that are not appealing to men whose basic operational mode is analytical.[29] Such terms as "current public issues," "international affairs," or even "current political issues" represent, on the other hand, aspects of life with which a scientist, as an intelligent person, cannot help being concerned. The level of this interest was assessed by a question that asked the respondent to estimate how often he discussed current political issues with his colleagues. Overall, only 2% replied "never."[30]

The level of the scientists' emotional investment in political affairs is somewhat lower, however, than might be expected on the basis of their interest. While possessing a generally high information level on matters pertaining to political affairs, the scientists in international laboratories display a tendency to remain rather detached from everyday political trends and developments. Several factors are associated with this tendency. First, opportunities for direct participation in political affairs are quite restricted. Second, the amount of emotional energy which professionally oriented scientists and engineers invest in their own work limits the depth to which they may involve themselves in outside activities, especially those that are neither recreational nor related to family life.[31] Third, in the international situation, one suspects that low emotional involvement may be functional in conflict avoidance. Finally, it appears that, in their political thinking, these scientists tend to extrapolate the methods and viewpoints of science itself. Where things do not quite fit, most are unwilling to make a deep personal commitment.

The words of a British engineer from ESTEC are suggestive:

I think scientists in general feel that politics are a state that go on anyway. You cannot classify politics in the same way, for example, that you can classify the elements, and whilst one will have opinions, one could only influence things in a minor way.... So they have feelings, but perhaps they don't inflict them very widely on other people.

Politics consists of phenomena that often appear to be quite irrational and whose practice is not normally susceptible to the "neat" methods of science and technology. It is an uncomfortable entry into a world structured by the extrapolation of science. Socialized to thinking within the norms of the scientific method, unable to impose rational solutions on political problems, and sharing what they admit to be rather idealistic notions about world affairs, many of the respondents seemed to temper their very real interest in political and social affairs with a sense of detachment.[32]

Under these circumstances we find that political discussions among the scientists normally take the form of exchanges of information rather than emotional disputes. The respondents' own words again tell the story:

If we invite a couple of English people to our home, the discussion is often about differences between Italy and England....So if we talk about politics, you keep on saying "in Italy one does like that," and they say, "no, in England one does like this." This is the kind of discussion one often has.
—Italian scientist at CERN

Most of our discussions are concerned with describing certain political features, you see. It's not very often a real argument.
—German scientist at CERN

Well we have a little bit of argument, yes, but normally the discussions concern what peoples' points of view are.
—Italian engineer at ISPRA

Extreme views are seldom found, and, as will be seen shortly, there is a broad sharing of political orientation and views among these international scientists. This consensus, together with the respondents' sense of detachment, generally serves to maintain political discussions at a low emotional level.

Another by-product of the respondents' low emotional investment in political affairs is a general feeling among them that their informational level is lower than it should be. Although we have no objective measure of this variable, it is our impression (based on the interview and questionnaire responses) that the level compares reasonably to other highly educated elite segments of society.[33] However, since they are trained as scientists, the respondents hesitate to draw inferences and propose firm conclusions on certain political questions, in particular those that seem to require expertise outside of their own specialty.[34] Accustomed to speaking with authority in their own fields, the respondents were unwilling to give "amateurish" opinions in areas where they felt less knowledgeable.

Interest and the Political Spectrum

The extrapolation of the scientific viewpoint to political thought leads to some readily identifiable trends among the scientists' attitudes. Within the conventionally recognized spectrum of political thought ranging from extreme left to extreme right, the largest part of the respondents found themselves in a moderate left-to-center position. The scientists were asked to define their own places on such a spectrum. Among the 237 respondents from whom there are data (the question was omitted from the interviews), 1% rate themselves as extreme left, 41% as moderate left, 36% as center, 21% as moderate right, and less than 1% as extreme right.

There is a very interesting and unexpected relationship between political orientation, as measured by this scale, and degree of political interest measured by a separate question. It appears that right-leaning respondents consider themselves less interested in political affairs than their left and center colleagues.

Table 5. Political Interest by Political Orientation (%)

Interest:	Orientation		
	Left (n = 100)	Center (n = 86)	Right (n = 51)
Well above Average	9	7	2
Above Average	35	19	8
Average	49	60	68
Below Average	4	8	18
Well below Average	0	1	2

Table 5 displays this finding. Among those respondents ranking themselves as leftists, 44% believe that they are above average in political interest. This figure drops to 27% among those in the center and 10% among the rightists. The cross-tabulation of political orientation by frequency of political discussion follows a similar pattern.

One of the central findings of American voting research is that of the positive association between partisanship and political interest. In American terms, this has meant that voters who considered themselves either committed Democrats or committed Republicans are much more interested in political affairs (especially elections) than so-called Independents.[35] If one attempts to draw a parallel between American political partisanship and the political spectrum data on European scientists, one is struck by the contrast in the distribution of political interest. Several alternative explanations may be proposed: (1) It is conceivable that since the American political parties are said to be really nonideological coalitions and do not correspond to left and right in any meaningful manner, then this parallel is not really valid.[36] (2) On the other hand, these findings may not necessarily be characteristic of the scientists in this sample but rather of a larger difference between European and American political cultures. (3) It is also possible that the scientists are evaluating their own standing within the framework of a limited reference group—the scientific community —and the scale they are using is linearly shifted with respect to the larger population. Hence, the leftists would be fairly strong leftists in relation to the general political spectrum, the centrists would actually be moderate leftists, and the rightists would really belong in the (disinterested and uninvolved) center. This last explanation is reflected in the words of one British engineer at ESTEC in an unconscious and disarmingly innocent way:

I think scientists are like any other body of people—they vary tremendously. You get some people who are very, very left and you get people who are a good deal less left.

Regardless of the reasons which lie behind the relationships between degree of interest and position on the political spectrum, it does appear that the leftist position represents something of a political norm in these communities. Since those who espouse it tend to show the greatest interest, they might be expected to be more influential as well. Just like the British engineer, who found some scientists "very, very left" and others "a good deal less left," many of the respondents observed the tendency toward a liberal-left consensus among their colleagues. The absence of extreme views and the generally leftist consensus are borne out by most of our data. The fact of being a scientist compounded by the situational factor of being a staff member in an international laboratory produces a rather narrow spectrum of thought concerning many crucial public issues. Although in the balance of this study we shall try to learn from such dissensus as we are able to uncover, it is important not to lose sight of the degree of unanimity with which most of the political questions were answered.

A tendency toward the left, however, does not indicate adherence to something that could rightly be called an ideology, in the sense of a codified doctrine associated with a particular party or political movement.[37] It is rather a shared set of values that produces a common political sentiment of leftism. This is what the French, long-habituated to such nuances, call gauchisme, and the scientists tend to be gauchisants (left-oriented) rather than gauchistes (left-affiliated).

From this orientation derive two sets of distinctive, though diffuse, images associated with the left and the right. In the minds of most of these scientists the image of the right is rather negative: its properties are self-interest, power, militarism, traditionalism, and pessimism. In contrast, those properties associated with "leftism" tend to be idealism, generosity, objectivity, and optimism—adding up to a much more positive picture.

These generalized sentiments create a broad transnational consensus that enables European scientists of diverse origins to feel "at home" with one another in facing the world political arena. It is not a "party line" in any narrow sense. It is rather a sharing of values and assumptions, style of thought, and tone of voice. A few excerpts from interviews done at CERN (in 1965) will convey the flavor of these images. No direct question was asked here; all of these responses are spontaneous, usually by way of explanation of each respondent's classification of his own political orientation. Here, then, are the words of four physicists of different nationalities.

If a person is on the right, he has some interests. If he is on the left, he has some ideals....In other words, he is capable of an objective evaluation of the world's social problems without taking his own particular interests into the account.
—Italian physicist

....Right people are fundamentally for themselves, for their own particular group. ...Whereas the left—they think of the maximum good for the maximum number.
—British physicist

....Rightists [in Germany] are very much for good old tradition; they are very much for good old allegiance; they are very much for good old soldiers and so on. I don't like it so much.
—German physicist

We can define two kinds of people. Those from the left think that generosity in other people is the main thing. One the other hand, the people from the right think that power is the main thing....What I call a left man is a little bit optimistic about the possibility....for man to do better—not in a trivial way, you see, but to a better knowledge of nature and man. On the other hand, a pessimistic view of life is held by the man from the right.
—Swiss physicist

It is important to emphasize again that the prevalence of leftist sentiments is the result of a preferred general orientation toward social problems and not the choice of one political party line or economic ideology over another. The notion of ideology, with its implications of rigidly codified "positions" on a wide range of issues, was roundly rejected by most of these scientists. Although they consider themselves "liberals," the scientists strongly reject any effort to impose a fixed pattern or structure on their views. They claim to prefer compromise, problem solving, and what seems to be a generally pragmatic approach to the political problems of the world.

Within the broadly leftist consensus, which one of the quoted respondents related to the internationalism of science itself, the not unexpected tendency of the respondents was toward a much stronger interest in international issues than in local or national ones. To these scientists, local and even most national issues seem trivial in comparison to the implications of international issues. Furthermore, to most, interest and participation in the routine activities of politics on the local level—attending party meetings and rallies, campaigning, fund-raising, and so forth—are clearly unappealing prospects. Finally, and perhaps most compelling, there is a need for a common ground in political discussions between scientists of various nationalities.

In the words of an Italian physicist from CERN:

We normally talk not about politics here but something different. We talk of all the problems. Not politics normally, but international politics—the trend of the world, not a question that is particular to a nation or something like that.

Participation and the International Life

As the scientists are employed on the basis of supranational professionalism, the well-being of the organizations requires that they remain aloof from activities serving private, partisan, or national interests. Each of the organizations hence imposes such a restriction as part of its employee regulations. In this respect, the situation is similar to that in other nonscientific international bodies and reflects the staff's privileged status in the host country. The staff members are well aware of these rules and apparently adherence to them is quite general.

The most common forms of political participation are activities that fulfill informational needs, and these often take the shape of public forums and discussion groups concerned with current international issues. For example the EURATOM centers, including ISPRA, quite naturally entertain speakers on issues of European integration. The staff association at CERN regularly invites speakers on public issues as part of its cultural program. During our visit, a lecture by a prominent Italian physicist (from outside the organization) reporting on a recent trip to North Vietnam drew a standing-room-only crowd in a large auditorium. An official from the American Consulate in Geneva was scheduled to speak on American Vietnam policy later that month.

On a different level, some of the top scientists involved in CERN, ESRO, and EURATOM have been among the participants in the Pugwash Conferences on Science and World Affairs. This type of activity, promoting an East-West dialogue, would probably appeal to many of the less-prominent scientists who comprise our sample, but the fact that they are not generally able to take part is more a function of their lower professional standing than their status as international civil servants.

FUNCTIONALISM AND EUROPEAN INTEGRATION

The portion of the survey which dealt directly with the political opinions of the scientists attempted to strike a balance between the extremes of overgenerality and over-specificity. It avoided, in other words, questions concerning highly current topics as well as those on the historical-philosophical plane. It was hoped thereby to uncover distributions of opinion on those issues most salient to the respondents, respondents whose political discussions, in the analogy of one ISPRA Frenchman, resembled "more those you can find in a weekly magazine than in a daily newspaper."

Despite their lack of topicality, the relevant questions were by no means immune to the winds of change, even over the relatively short term. The European milieu of April-July 1967, within which the scientists framed and evaluated a range of policy choices, differed substantially from that of today. Throughout this discussion the reader ought to bear this fact in mind.

A further caveat also bears mention. Our approach to the treatment of the political opinion data is analytical and to an extent quantitative but makes no pretense at definitive quantification. We feel that this is appropriate in view of the exploratory nature of the study and our limited objectives in dealing with this material. In essence, what was done after delimiting an issue cluster was to define the range of views observed on facets of this cluster, establish the consensual view (to the extent it exists), examine the deviations, relate views on various facets of the issue cluster to one another, and look for the differences produced by crosscuts of the prime independent variables (laboratory, nationality, and seniority) as well as other variables. We rely largely on marginals and cross-tabulations of quantitative data while drawing freely from the richer interview texts and write-in comments.

Economic and Political Integration

Support for common economic institutions that have developed in postwar Europe is so widespread among the elites in the major nations that "favoring" these institutions is no longer a relevant choice. In their report on the TEEPS project, Lerner and Gorden write:

The success of the appeal [of collective prosperity] to the elite attitudinal base was so clearly consensual that, by 1961, we stopped asking panelists about the European economic organizations. The consensus that had been built since the early years of the TEEPS surveys became so strong that the idea of collective economic growth had ceased to be a matter of discussion. [38]

We felt sufficiently confident that this assertion should be valid for scientists and engineers working in joint European organizations to take it more or less as a base and go on from there in our questioning. The choices that were examined thus concerned the geographic extension of the European economic institutions as well as the potential movement of integration from the economic to the political realm.

Attitudes toward the problem of expanding the European Economic Community, choosing between a "little Europe" of six nations and a "big Europe" of ten or more nations, were explored through the straightforward question: "Would you favor the formation of a wider European Economic Community that would include the present EEC (Six) and other Western European nations?" This choice was expected to be most crucial to EURATOM scientists whose personal fortunes were directly tied to the political structure of the European Community. Some nationality differences were also anticipated on the basis of differing national involvements in such a venture. In fact,

Table 6. Extension of the EEC in Western Europe, by Laboratory and Nationality (%)

	ESTEC				CERN				ISPRA		
	Br n = (35)[a]	Fr (19)	Ge (20)	It (7)	Br (25)	Fr (13)	Ge (17)	It (19)	Fr (25)	Ge (35)	It (39)
Yes	83	68	100	100	88	100	100	95	84	91	85
No	11	26	0	0	12	0	0	5	12	3	10
DK[b]	6	6	0	0	0	0	0	0	4	6	5

[a]These n's are identical for the remaining tables in this chapter (except where indicated) and will not be repeated.
[b]DK is used throughout to indicate "don't know," which includes "no response."

the degree of unanimity with which the scientists approved expansion of the EEC (90%) was rather impressive. Table 6 presents the nationality/laboratory cross-tabulation for this question. Out of a total of 384 respondents, only 26 (7%) disapproved of extending the EEC to other countries of Western Europe. Within this lopsided distribution lies, in one sense, an even stronger unanimity, but, in another sense, a definite ambivalence. The finer structure is revealed through analysis of write-in comments, interview transcripts, and cross-tabulation with other questions.

It is first of all evident that virtually all of the scientists, whether or not they are officially affiliated with the European Community (EURATOM), approve of the "big Europe" scenario.[39] Even among the 26 dissenters, more than a fourth qualified their negative responses by annotating comments which indicated that they would approve of extension at some time in the future. Further, it is apparent from individual examination of the questionnaires that about half of the remaining disapprovers were motivated in their disapproval by a desire to foster rapid political unification of the Six but were definitely interested in eventual expansion.

The fear that expanding the EEC might slow the process of political integration is at the root of an ambivalence that we suspect underlies the apparent unanimity of this response. Consider the following interview excerpts, all responses to the foregoing question:

In a way, yes. This means that I think it's necessary to reach a political union [in Europe] and I would like to see this union as big as possible. On the other hand, if there was a union between the Six and the Seven, but purely on an economic basis, then I would be against such a union.
—EURATOM Dutch physicist

These are difficult problems...it is quite difficult for different nations—even when there are not too many—to reach a common point of view. Now, if you take a really strong union and add more nations, you make things more difficult. I mean it may

turn out like the United Nations: people discuss and discuss and nothing comes of it.
—CERN French physicist

Basically, I would prefer a larger community, but again I have in mind this idea of
reaching as fast as possible a political union. Now, if the political union becomes
more difficult by getting other nations into the Common Market, if the Common Market would gradually develop into an organization of only economic cooperation and
assistance, then I would prefer to have only six nations.
—EURATOM German physicist

Our question was not phrased in terms that might force this hard choice between expansion of the economic community and political integration of the present grouping,
and perhaps it will not be necessary for Europe to decide between these alternatives,
but it appears that some misgivings of this nature underlie the nearly unanimous desire
among our respondents to create a "big Europe."[40]

Naturally, three of the four major nationalities were able to view the question of
expanding the community from a different perspective than the fourth. To the British,
the choice was between joining the EEC and remaining outside of a European union.
While the overwhelming majority of British respondents did approve of expanding the
community—by implication including Britain—it is noteworthy that the British accounted for a significant proportions of the disapprovers. The meaning of this British
response will become clear shortly.

The thrust of the preceding analysis suggests that it is necessary to look to the
matter of political unification for further definition of this opinion cluster. The questions "Are you in favor of the formation of a political union among the 'Six'?" and
"Would you favor a political union on a wider scale in Europe?" probe this matter.
Their tabulations are presented in Table 7.

Again, one may observe a strong consensual response approving the concept at
issue. Aside from the British and French at ESTEC, more than four-fifths and up to
100% of each group favor a political union of "little Europe" and nearly the same proportions favor wider political integration. Even among the deviating ESTEC respondents, there are healthy majorities in favor of political unification and relatively small
numbers directly opposed.

Such strong approval as was found at the large centers was not unexpected, of course.
Political integration, despite its apparent lack of progress in the realm of government
action, has met with favor among the elites of Western Europe for some time. Elite
opinion surveys in the major European nations give evidence of this trend. Deutsch
et al., for example, reported that in their 1964 study of elites in France and Germany, 67% of the French elite favored "some kind of 'supranational dominance'" in
European integration, and virtually all of the German elite did so as well.[41] The

Table 7. Approval of European Political Union, by Laboratory and Nationality (%)

	ESTEC				CERN				ISPRA			
	Br	Fr	Ge	It	Br	Fr	Ge	It	Fr	Ge	It	TOTAL
n =	(31)	(16)	(20)	(6)	(21)	(8)	(16)	(18)	(20)	(32)	(35)	(342)
Union of Six												
Yes	57	79	95	100	88	100	100	90	88	97	97	85
No	34	16	5	0	12	0	0	5	4	0	3	11
DK	9	5	0	0	0	0	0	5	8	3	0	4
Wider Union[a]												
Yes	61	56	90	83	86	87	100	94	85	91	83	81
No	32	2	5	0	14	0	0	6	10	3	11	12
DK	7	19	5	17	0	13	0	0	5	6	6	7

[a]The pretest interviews led the author to believe that the subject of wider political union would arise so naturally out of the discussion that it would not be necessary to include a separate question with regard to it. This expectation was not borne out in 42 of the later interviews. The number of those for whom data are available is given in parentheses, and the percentage for the "wider union" question is based on this number.

TEEPS study, report Lerner and Gorden, discovered similar attitudes in France and Germany but not in Britain:

The political interpretation of Europe was given by the continental panels alone; we can now see that the politics they desired were those of close European supranational cooperation. The British... did not have such an orientation in mind. [42]

The ambivalence that some of the respondents showed toward the expansion of a purely economic community is apparently based on a strong desire for rapid strengthening of political ties between the European nations; confirmation can be seen in the political union response. The phrases "building Europe," "making Europe," "when Europe is done," are clearly part of these scientists' lexicons. The concept arose naturally within the interviews, not only with respect to the specific questions with which we are dealing here but in many other contexts. Only a few individuals did not speak in this vocabulary or were not quite at ease with it. Indeed, paralleling the TEEPS findings, these individuals were primarily British; most were from ESTEC.

The cross-tabulation of the two political union questions, distributed by laboratory and nationality, bespeaks this conclusion even more vividly. Here (Table 8), it is clear that, outside of the ESTEC British, the number of respondents who approve of

Table 8. Approval of Both, One, or Neither Form of Political Union, by Laboratory and Nationality (%)

	ESTEC				CERN				ISPRA		
	Br	Fr	Ge	It	Br	Fr	Ge	It	Fr	Ge	It
n =	(29)	(15)	(20)	(6)	(21)	(8)	(16)	(18)	(19)	(31)	(35)[a]
Approve Both	52	47	85	83	81	87	100	83	90	94	83
Only Six	7	33	10	17	5	13	0	6	5	6	14
Only Wider	14	13	5	0	5	0	0	11	0	0	0
Neither	27	7	0	0	9	0	0	0	5	0	3

[a]Excludes those who were not asked about wider union and those DK on both parts.

neither form of political union for Europe is miniscule. Small numbers of respondents are unwilling to see a political union of only six countries; others see wider political integration as unrealistic but most evidently do not advocate one form to the exclusion of the other.[43]

There is a distinct consensual position on the issue-cluster described so far. Over the entire sample, some 85% of the respondents approve both geographic expansion of the European Community and some form of political union.[44] These 85% are the "Europeans"; they hope that the political boundaries separating the existing nations will someday give way to a supranational community and they realize that, in the words of one interviewee, "the Six is not Europe, yet, it is just Six." They differ among themselves as to means, being undecided about whether to seek a political community on a smaller scale first and then expand geographically or to enlarge the community first as an economic community. The data do not allow us to distinguish various types according to their schedule of priorities, but we feel intuitively that this is not a crucial failing, as it seems that the respondents themselves are quite undogmatic on this point and most would be willing to follow whichever course appears more pragmatic for the goal of "making Europe."

The deviates from this consensual position are found in several categories: the British (13 in number) and the French (4) at ESTEC, the French (5) and Italians (6) at ISPRA. The ESTEC and ISPRA French and the ISPRA Italian deviates are relatively few, comprise 20% or less of their groups, and seem to be, for the most part, Europeans of a more limited sort. Excepting a handful, they are the ones who are most strongly against expanding the economic community on the grounds that it would weaken the ties between the Six nations. Perhaps because of this still European

orientation or perhaps simply because they are so few in number, it is not possible
to distinguish them from their colleagues on the basis of other variables. The ESTEC
British deviates, on the other hand, definitely appear to have less enthusiasm for
European supranationalism both in the economic and political senses; in the conventional parlance, they are less "European." The tone of this attitude appears, for
example, in one Briton's evaluation of the importance of the Economic Community:

There is this removal of some tax barriers but, again, not a great deal has been
done in this way. These Six work together in that their tariffs on some things are a
little bit lower than with the rest of Europe, but it's not terribly significant.

Let us make no mistake. The majority of the British scientists surveyed are
clearly European, committed to seeing their nation part of a supranational Europe.
At CERN and at the smaller laboratories, the percentage of British falling outside
of the consensus is about the same as that of any other major nationality. In this
respect, the views of the British scientists differ significantly from those of their
compatriots in the national panels of TEEPS. At ESTEC as well, the majority of the
British (63%) are also within the consensus. Those whom we have termed "deviates,"
however, are of particular interest to us here for they possess several distinguishing characteristics. They appear to be less interested in political affairs than the
other ESTEC British: none of the deviates ranks himself above average, while 23%
of the consensus group do so. They discuss politics less frequently: only 16% respond
"very often" or "often" versus 32% of the consensus group. They also consider themselves more on the right: 9% of the consensual group place themselves on the right
on the political spectrum while 38% of the deviates do so. [45] Finally, they have spent
less time in the international situation: only 23% of the deviates have spent more than
two and one-half years in the laboratory, while nearly 40% of the consensual group
have done so.

This last point leads to one of the few "tests" of functionalist theory which can be
made with these data and thus requires a bit of digression. Initially it was hypothe-
sized that through experience in an international laboratory, scientists would develop
attitudes increasingly favorable to European unity, thus preparing themselves for a
larger role in the integration process. A real test of this hypothesis, which would
have been highly desirable, was not permitted by these data. For one thing, characteristics inherent in the structures of two of the three large laboratories (ESTEC and
ISPRA) made it impossible to obtain a balanced seniority distribution. Beyond this,
there is the more fundamental matter of homogeneity of political opinion. On the
matter of European integration, some 85% of the respondents—17 out of every 20—

fit into the consensual position outlined earlier. With such global agreement, it is impossible to infer subtle shifts in attitude; there are simply too few non-European-ist respondents to allow one to see any differences when "slicing" by seniority.

This overwhelming consensus may be viewed at different levels of explanation. It is possible that the choices presented by the questionnaire were not really the crucial ones to the respondents—that is, the questions were "too easy" for this set of re-spondents. Looking at things in this light, one may hypothesize that changes in opinion patterns might still occur as a function of international experience, but that such changes are too subtle to be detected by the questions that were posed here. Alter-natively, one might simply take the data at face value: the bulk of the scientists are highly Europeanist when they come to the laboratories, and they remain that way.

This is where the ESTEC British become interesting. Within this group, those individuals with long-term experience in the laboratory are significantly more Eu-ropeanist than their recently arrived colleagues. With respect to professional vari-ables (field of training, type of work, level of highest degree, and reasons for coming to the organization) as well as such variables as age and prior foreign experience, the consensual and deviate groups with high and low seniority are reasonably homo-geneous (the numbers of course become quite small). In the absence of other "ex-plaining" variables this leaves us with at least a clue to the effect that the experience of working in a European international laboratory may make a British scientist more "European" in his political outlook. [46] The British at ESTEC, farthest removed from their colleagues in original political outlook, may be the ones most affected by the experience.

Expectations

It is of limited value to ask abstractly whether a European scientist would favor po-litical integration of the European states; a person's preferences in his "best of all possible worlds" often become unrecognizable when extruded through the mold of political reality. It was considered important, therefore, to determine whether the scientists were simply paying lip service to the ideal of a united Europe, or whether they had a concrete notion of its coming about.

The question is already partially answered. The tone in which the respondents em-ployed such phrases as "building Europe," "making Europe," and "when Europe is done," seemed to imply a genuine expectation on their part. Nonetheless, additional light is shed on this issue cluster by analysis of several direct questions bearing on the subject.

Political integration of the Six is viewed as a realistic prospect by more than two-thirds of the scientists. ("Do you think it will actually occur?" 67%, yes; 24%, no; 10%, don't know.) There is, furthermore, a high degree of interaction between preferences and expectations. Among the 325 respondents who favor political unification, more than three-fourths (77%) feel that it will occur, while only 16% do not. Among the 44 who stand opposed to a political union, 86% state that it will not occur, while only 11% feel that it will. Clearly, those who favor a political union see it as a realistic prospect. Some among the minority of respondents who favor political union but responded negatively (or don't know) to the question of expectation, furthermore, may have been giving vent to an expression of momentary frustration as much as anything else; more than a fifth still responded to the following question and estimated how many years it will take, while a number of others qualified their "no" by writing in "not for a long time."

The estimation of a time scale for political integration of the Six is a matter of some uncertainty among the respondents and no real consensus exists. The only period clearly ruled out by the scientists is the next five years. Among those who felt it will occur, only 1% estimated "less than five years," while 16% chose "5 to 10 years," 34% chose "10 to 20 years," 32% chose "more than 20 years," and 18% did not know. [47] Variations in this response by laboratory/nationality groups were irregular, but it did appear that the ISPRA scientists envisioned a somewhat shorter time scale than the others: 46% of them estimated twenty years or less, while 33% at CERN and only 29% at ESTEC made comparable responses.

While ten to twenty years may be a relatively short period in a historical perspective, it evidently seems inordinately long for many of these respondents in view of their strong desires for European unification. A number of typical interviews (which have been reproduced here conversationally) may illustrate this point. First a Belgian engineer at ISPRA:

A: . . . I have the feeling that political Europe is not made and it will take still many, many years before we arrive at such a concept.

Q: What kind of time scale?

A: I have a feeling of ten years, something like that.

Next, an Italian engineer from ESTEC:

Q: Do you think that the Six will eventually form some kind of political union?

A: Yes, but not as early as they say. I think that they will be obliged to do this, but it will be very difficult.

Q: Do you have any idea how far in the future this might be?

A: Oh, at least ten years.

Finally, a German scientists at ESDAC:

Q: Do you believe that the Six will eventually form a political union?

A: Well, I hope so ....It is not possible to achieve a political unification in ten years, you see. We have a history of 2,000 years and each country has its own history and such individualistic people as we have in Europe—Frenchmen, Englishmen, and Germans—won't find it easy to do this....

Q: Well, how many years do you estimate it might take?

A: Oh, I would say twenty, thirty years.

Clearly, the respondents anticipate unification of at least "little Europe" in the relatively short-term future. Integration on a larger scale is a different matter, however. Only a minority (31%) indicated that they thought wider political union was possible in the foreseeable future; more than half (56%) did not foresee such a pos-sibility, and 13% gave no opinion.[48] There was some interaction between the ex-pectation of union of the Six and that of a larger union. Of those who stated that a union of the Six would not occur, only 13% anticipated the broader union (82% said "no" and 5% did not know). Among those whose expectations did include a union of the Six, nearly half also felt a wider union was possible.

Integrative Motives

What hopes do the scientists, so committed to the uniting of Europe, place in the ac-complishment of this aim? Other studies have attributed various motives to the elites of the major European nations in their quest for integration. Deutsch et al. report the most popular French view of the purpose of European integration as "generally, to give Europeans the means and resources to solve economic, social and political problems currently insoluble by nation-states acting alone." "Economic betterment," "reinforce[ment] of the European bargaining position in world politics," and "strength-en[ing] the European position vis-à-vis the United States," follow in that order.[49] Unification for Germans is seen as all things to all men—economic benefits are stressed by businessmen, political benefits (inside Europe) by politicians, and dip-lomatic benefits (outside of Europe) by administrative and communications elites.[50]

Lerner and Gorden propose a paradigm consisting of three different policy prior-ities, "protection, prosperity, and prestige," which are differently ordered under Euronational, Euramerican, and Euratlantic" scenarios. (The "Euronational" scenario corresponds roughly to the Gaullist design; "Euramerican" to the Monnet model, and "Euratlantic" to such initiatives as NATO and the Marshall Plan.) They conclude that

there is an emerging consensus within, and a convergent consensus between, the elites of the European nations. Further, that this convergence goes in the direction postulated by our developmental construct: from nationalism to regionalism. The convergence is based on the shared long-run expectations that personal and public values (protection, prosperity, prestige) will be enhanced by the larger community of interest embodied in the Euramerican and Euratlantic scenarios.[51]

While this is a rather general statement of motives, it corresponds well to the overall frame within which our respondents see unification operating. To a large extent it appears that, within the consensual position of these scientists, the sustaining value of integration is so automatically taken for granted that its specific benefits are not of terribly high salience. Among those who share the "European" orientation, the purpose of "Europe" is beyond question.

We did not attack this facet of the unification issue cluster directly, but found that the relevant data emerged within the context of the interviews, mainly in the analysis of three questions. While the three are taken up here in varying depth, they all share the aspect of being concerned with Euro-American relations: the first with an Atlantic Economic Community, the second with "independence," and the third with the "technological gap." What emerges from their analysis is a feeling that a best approximation to the respondent's motivations for European unification is implied in the words of one scientist, "If there is a future for Europe as a leading continent, it's only in unity." By this is meant in Deutsch's terms "generally giving Europeans the means and resources to solve economic, social and political problems currently insoluble by nation-states acting alone," and in Lerner's terms the enhancement of the public values "protection, prosperity, prestige." Above all, though, there is the impression that particular reasons are not so important as the broad vision of a United Europe.

This vision may be seen in terms of Euro-American relations. Following the question on expansion of the EEC in Western Europe were two additional expansion proposals, the first suggesting an economic community with the nations of Eastern Europe, and second suggesting such a community with the United States. Close to two-thirds (62%) of the respondents approved expansion to Eastern Europe, while fewer than one-third (31%) approved expansion to the United States. The latent reason behind this response appears clear: Eastern Europe is "Europe" and the United States is not.[52] While "strengthening the European position vis-à-vis the United States" is not necessarily a primary motivation for these Europeanists, it is certainly a well-perceived outcome of the process. Such thinking is by no means anti-U.S. to the scientists; it is merely pro-Europe.

The relevant portion of this question (which read "Would you favor such an economic community including the United States?") drew an extraordinarily high rate of write-in comments (half of all respondents qualified their answers with volunteered remarks) and proper interpretation requires taking these into account. Most impressive is the fact that both "yes" and "no" respondents made very similar remarks and qualifications. The bulk of the comments reflected feelings that the purpose of economic integration is the building of <u>Europe</u>, and that the time to talk of stronger Atlantic ties was after Europe was united. Often the fact that the United States is so much larger and more powerful than the individual European countries was cited, and it was explained that unless Europe was united first, the inevitable result would be American domination of any integrative arrangement. Many comments on both "yes" and "no" responses deferred consideration of the matter to the future, and one is led to believe that, while there are variants from this regionally oriented position (for example, feeling as one Englishman did that an Atlantic community would divide the world even more rigidly between the "haves" and the "have-nots"), it is a more broadly shared notion than one might conclude from the marginal alone.

A few interview excerpts may also be illuminating. The words of a British scientist from ESDAC suggest the shape of the position:

Here I think one has to say why one is interested in extending the Common Market, and it's not simply for economic reasons. It's more for the fact that this gives one sort of a foothold in international cooperation. Well, in Europe, generally, Western Europe, this sort of thing exists to some extent already. Eastern Europe, one has no objections, but it's a much more remote prospect, you know. This could be integrated very readily, given the right political circumstances, into a viable economic unit and a sensible political unit. With the United States, I'm not so sure that this is true to anything like the same extent.

Those of a French physicist from CERN refine it:

I would be a bit worried by the fact that there is so much difference presently between the weight of the United States and that of Europe which is not at all unified, strongly unified. All the weight would go in one direction and I don't think it's at all the time for this now.

Finally, a Belgian engineer from ESTEC may have put it most concisely:

This is a funny question. The United States mustn't be included because nothing would be changed in this case!

It is by and large a matter of independence. A unified Europe, regaining its position as a "leading continent," must be able to determine its own destiny. When asked if they thought Europe "should take a more independent position in the Western Alliance," nearly three-fourths (71%) of our respondents replied affirmatively. This response, which again was shared across the various laboratories and nationalities, is further supported by the fact that a significant proportion of the negative responses were qualified by such comments as "after it is united." Even among those respondents who would approve an Atlantic economic community, a majority (55%) favor a more independent position for Europe. Here, then, is the "prestige" element of the motivation for building Europe.

Finally, as scientists and engineers, it is perhaps in their own fields that the need to build Europe strikes our respondents closest to home. When asked if they thought "that the 'technological gap' currently being discussed in the press is a serious problem for Europe," the vast majority of the respondents (75%) replied affirmatively. The follow-up to this question was open-ended and asked the respondent to suggest whatever action might be necessary to close the gap. Responses ranging from short catch phrases to lengthy proposals were obtained, and by far the most frequent were those that emphasized more European cooperation and integration. Nearly half (42%) of all the responses took this form. Drawing on the interviews again, one may cite answers such as:

No, I have no solution for this problem. . . but I have the feeling that making Europe is something which is really necessary, which is a fundamental necessity to arrive at a solution.
—EURATOM Dutch physicist

Well, I feel that the only type of action that could have a chance to fill that gap would be an action of the Community. I believe if we don't build Europe, at least technological Europe, we have no chance.
—EURATOM Belgian Engineer

Essential to remember here is the fact that Lerner and Gorden's so-called "Euramerican Scenario" emerges very clearly from a significant proportion of the respondents of all major nationalities without any attempt at direct elicitation.

The Shape of a United Europe
Beyond the outlines of European unification, which were drawn first in terms of preference and expectations, and second in terms of the hopes vested in the process, the images of actual structure held by the international scientists and engineers become somewhat more diffuse. In view of the fact that they do not possess the technical skills necessary for sophisticated examination of many of the questions involved and in view

of their limited emotional commitment to political matters, such "thinning-out" should not be unexpected. Nevertheless, the survey had moderate success in dealing with two segments of the structural complex: first, with the locus of power in a United Europe, and second, with the preservation of national cultures.

The preservation of national characteristics might be expected to be a source of ambivalence toward European integration: Is there a perceived conflict in the minds of the respondents between the desire for prosperity, protection, and prestige through the elimination of national boundaries and the desire for maintaining the special char-acter of Europe--the individuality of its parts—of which Europeans are so fiercely proud? On the other side, the particular form of government for the Europe of the future, while perhaps a matter which the respondents would regard as outside of their competence, will be one of the "hard" choices that the Europeans will have to face in their quest for unity. As part of the debate over "Europe des patries," it has already been the source of controversy between Gaullists and more supranationally oriented Europeans.

Nothing that even remotely resembles a consensus came out of our question on form of government. The proportions favoring a "strong central power" (34%) versus "more power for the states" (47%), give a slight edge to the states. But, a large DK (17%), wide variations between the various laboratory/nationality groups (see Table 9), and an ambiguous set of volunteered comments apparently make it a wide-open issue.

The question itself and the choices proffered were apparently themselves ambig-uous since there are widely differing conceptions among the respondents of what con-stitutes strong or weak central power. This makes it impossible to base any firm conclusions on a respondent's choice of one form of government over the other. One may note, for example, that, in write-in comments, 12 respondents took the United States as an instance of a strong central power, while 10 took it as an instance of a form with more power for the states. Strong central power, to the extent it exists in France (or Italy), is quite different from the American conception, and it is likely that few respondents envisioned anything quite so centralized for a European govern-ment. [53] Probably the most widely shared view, to the extent that we are able to un-cover it, incorporates a federal structure roughly paralleling that in the United States, with certain powers delegated to a central authority but with substantial local auton-omy.

The second aspect of the image of a united Europe is that which must rationalize a desire for political integration with a desire to maintain cultural differentiation. To

Table 9. Preferred Form of Government, by Laboratory and Nationality (%)

	ESTEC				CERN				ISPRA		
	Br	Fr	Ge	It	Br	Fr	Ge	It	Fr	Ge	It
Central Power	31	53	40	14	20	31	35	26	20	40	54
More for States	40	32	45	43	64	23	59	48	48	34	36
DK	29	17	15	43	16	46	6	26	32	26	10

see if our respondents had succeeded in resolving this difficulty, we asked them, "Do you think that the existing cultures of Europe could maintain their individuality within a united Europe?" The response was a resounding "yes" (88% "yes" versus 6% "no" and 6% "DK") and the consensus covered all laboratory/nationality groups. Only a handful of respondents found any problem at all. A great many pointed to the numerous contemporary European examples which demonstrate that there need not be identity between culture and political structure:

Look at Switzerland, for instance. This is a nation-state but the provincial character-istics are still there. Everyone still has his own language, his own culture and so forth. We have a very good example in Europe which we can follow.
—EURATOM Italian engineer

I mean the Bavarians still maintain their individuality in Germany and the North Ger-mans are certainly quite different from them. So there is no difficulty.
—CERN German physicist

There is no desire on the part of the respondents to see a blending of the various na-tional traditions, and there is no reason, in their minds, why political integration should lead to this. There is even a mild suggestion that European integration might facilitate the preservation of national traditions by resisting the tide of "American-ization."[54]

MILITARY ISSUES

Although issues of European integration are most salient to these international scien-tists, such questions do not demarcate their political horizons. Of particular concern, because of their far-reaching consequences, are problems that involve East-West re-lations, military security, and nuclear weapons. The data permit us to gain some very interesting insights into how the scientists think on these matters.

East-West Relations

The whole notion of détente between East and West is so widely taken for granted within the scientific communities of the various Western European countries that we felt it would be superfluous to ask respondents if they favored the concept. [55] The survey found no evidence to contradict this assumption. In fact, as has already been noted in the previous section, the possibility of opening up an economic community between Eastern and Western Europe was welcomed by a sizable majority of the respondents (62%). [56] This kind of step goes well beyond the current notions of deténte. Thus, the question which was asked on this topic dealt with expectations rather than preferences and was phrased in the following manner: "Do you feel that the détente which has been developing in recent years between the West and the European Communist countries is part of a lasting trend?" The response was an overwhelming affirmation (75% "yes," 8% "no," 17% "DK"), and comments such as "I hope so" followed nearly every positive interview reply. Table 10 shows that the unanimity of this response is among the most widespread of those with which we have dealt.

Expectations that the détente is part of a lasting trend range from 64% to 100%, and in those groups where optimism is weakest it is replaced not by pessimism but rather by uncertainty. About one-quarter of the respondents (if one may extrapolate motives from the interviews to the entire sample) were simply unwilling to state categorically that their hopes would be fulfilled in the foreseeable future; most chose to say that they did not know, while a few expressed the equivalent of "no, there are too many fluctuations in East-West relations to be able to project any lasting trends."

The expectation of détente transcends the division of the sample into laboratory/ nationality groups. Furthermore, inspection reveals no consistent relation between the response patterns of these groups on the question at hand and on the question of an economic community including the East. Two points should be noted, however, with respect to the détente question. First, it appears that the scientists are not deeply

Table 10. Expectation of Détente, by Laboratory and Nationality (%)

	ESTEC				CERN				ISPRA		
	Br	Fr	Ge	It	Br	Fr	Ge	It	Fr	Ge	It
Yes	74	84	70	100	80	77	65	74	80	66	64
No	14	0	5	0	12	0	6	5	0	11	15
Dk	12	16	25	0	8	23	29	21	20	23	20

involved in this issue, at least in comparison to the issue of European integration.
In the interview situation, the question on détente provoked mainly platitudinous
responses. Second, although expectations of détente reflect a definitely optimistic
frame of mind with respect to European or even U.S.-Soviet relations, for at least
some of the respondents this evaluation was based on a fear of a growing threat from
a common enemy—China. Thus, one must be careful not to project such optimism
beyond the immediate stage within which it was expressed.

The atmosphere of détente, even qualified by some vague apprehension over the
future role of China in the international power balance, is consistent with a general
impression of other elite and public-opinion studies that indicate that Europe does
not feel militarily threatened. Such feelings go back at least as far as the mid-1950s,
as Lerner and Gorden found:

Consistently, throughout the decade, the fear that the Soviet challenge was primarily
military was discounted in favor of a predominantly political evaluation. Thus, while
the challenge was perceived throughout the decade ... it was never evaluated primarily in military terms by any panel. [57]

The absence of a direct military threat, coexistent with a strong alliance involving
the nuclear capability of the United States (to counter unanticipated military pressures),
makes the military component of European integration a relatively low priority subject. In discussions of European integration, in fact, the military aspect did not arise
spontaneously in a single interview. It was evidently not highly salient either as a
motivating force for integration or as an outcome of it. In this respect, the scientists
do not differ from the vast majority of other European elites whose vision of European
integration Deutsch characterizes as "primarily non-military in purpose."[58] This is
not to say that European military integration was not discussed at all in the interviews.
Rather, the whole concept of integrating national armed forces into supranational
forces had to be raised independently, and in this context, Europe was posed as one
of a range of options.

Attitudes toward the general concept—"Would you approve the integration of a major
part of your own country's armed forces into a permanent supranational force?"—
were ascertained first. Then, a choice of auspices was presented: Europe, NATO,
and the United Nations. Although a broad concensus dominated the responses to the
first part of the question, one clear national difference, consistent across laboratory
lines, is evident, and more differences crop up in the second (auspices) part of the
question. Overall, some 78% of the respondents approved the abstract concept of integration of their nation's armed forces into some supranational structure. Only 11%

Table 11. Approval and Auspices of Supranational Force, by Laboratory and Nationality (%)

	ESTEC				CERN				ISPRA		
	Br	Fr	Ge	It	Br	Fr	Ge	It	Fr	Ge	It
Approval											
Yes	80	53	85	100	84	54	100	84	60	97	77
No	11	21	10	0	12	23	0	11	24	3	8
DK	9	26	5	0	4	23	0	5	16	0	15
Auspices, if Such a Force Were Created											
European	26	53	45	43	12	54	41	37	56	60	51
NATO	17	0	20	29	16	0	18	0	0	14	10
UN	37	0	15	29	64	8	18	26	8	20	15
Other, DK	20	47	20	0	8	38	24	37	36	6	23

opposed integration and an additional 11% gave no opinion. Table 11 presents the laboratory/nationality distribution of this question as well as the distribution of the subsequent question on auspices.

The clarity and significance of the national divergence on the first part of this problem is beyond question. The French respondents, in sharp contrast to respondents of all other nationalities, show a level of support just barely over a majority. French disapproval is double that of the sample as a whole, while the French DK is also well above all other groups. Further, the response is consistent across the three major laboratories. While it is, of course, important that the majority of French scientists go along with the majorities of scientists of other nationalities, it might be instructive to attempt an interpretation of the reasons behind the weakness of their approval.

Such an interpretation comes directly out of examination of the second half of the question. The three options proposed represent widely differing notions. An integrated NATO force is an entirely different type of animal than a European force, while an integrated United Nations force differs even more from these two than they do from each other. In virtually all cases except the French, however, approval for all three types of supranational auspices is evident. Only among the French is approval very tightly drawn: if a French scientist approved of a supranational force, he had in mind specifically a <u>European</u> force. It is proposed that two main sorts of attitudes governed the responses to this question, and although they are not mutually exclusive, one or the other tended to show up in a given respondent. On one hand there is the antimilitary attitude. For many respondents who held this posture, the particular auspices of integration are not as important as the primary aim of eliminating national armies. Responses such as the following, from an Italian scientist at ISPRA, typify the attitude:

A: Yes, I would approve [of military integration]. The idea is to eliminate them, so I have no difficulty in approving it.

Q: Under what auspices would you prefer to see it: European, NATO, or United Nations?

A: It's important only that we have an international organization. It is not important which is the organization, only that it is really international.

On the other hand, there is the attitude that views favorably an integrated military force within the structure of a United Europe. Importantly, many respondents mentioned that they would prefer to see the formation of such a force <u>follow</u> the political unification of Europe rather than <u>precede</u> it, since, in the words of one Briton, "sufficiently strong political leadership must come first." In this attitude, the supranational force is merely one somewhat secondary aspect of an integrated Europe.

The two attitudes (and/or their admixture since, we repeat, they are not mutually exclusive) were present in all of the laboratory/nationality groups in our sample. Among the French, however, the antimilitary response apparently manifested itself in a rejection of the supranational force concept rather than in approval without strong specification of auspices. Consider, for instance, the response of a CERN French scientist:

What I am for is the suppression of all military forces, so I don't think this is any good. I would rather suppress all of them, rather than try to make a supranational one.

A characteristic French trait of rejecting questions seen as too vague,[59] combined with a lower French evaluation of the United Nations and NATO as effective international organizations, produced this response pattern. Where the European attitude was dominant, approval of the supranational force with specifically European auspices resulted.

Overall, the qualitative impression that one received from the interviews, is that the scientists' views on the subject of a supranational military force were not nearly as highly developed as their views on European integration. More vagueness and generality in phrasing, as well as longer pauses, were taken as indications that the

subjects had not thought about the questions in great detail. Further, the scientists
seemed less willing to get involved in this whole area of discussion. As few real
proposals for an integrated European force (outside of a political union) had arisen
since the defeat of EDC in 1954, and, as most Europeans never seemed to develop
real enthusiasm for the MLF-ANF concept, the lack of saliency of this question may
come as no great shock. The lack of real motivation for developing large armed
forces—we recall the absence of a sense of military threat mentioned earlier—and
the antimilitary bent of many scientists no doubt also contributed heavily. It was
more than an isolated interviewee who, when asked if he approved integration of his
country's army into a supranational force, retorted at first, "For what? To fight
against whom?"

Nuclear Weapons: National and Supranational

In light of the range of national situations, wide variations in response among scientists of British, French, German, and Italian nationality should be expected on questions dealing with national nuclear policy. The question "As a matter of general
policy would you be in favor of possession of thermonuclear weapons by your own
country?" was, in fact, the only one in the entire questionnaire which drew a strong
reversal across nationalities. While the French and British scientists were split
around the 50% mark, with a majority (54%) and a plurality (46%), respectively, in
favor, the Germans and Italians showed overwhelming majorities against nuclear
weapons with only a very few (8%, 13%) in favor. Table 12 presents the distribution
of this response by laboratory and nationality.

The national situations—controversiality in Britain and France and strong disapproval in Germany and Italy—are echoed in the response patterns across all of the
laboratories. It is noteworthy that opposition to the national nuclear forces among the
British and French is stronger at CERN than at ESTEC and ISPRA. Although all three
of the laboratories are concerned exclusively with nonmilitary, peaceful research, the
proportions of British and French scientists and engineers who have been involved in
military and atomic energy research are higher at ESTEC and ISPRA than at CERN.
CERN, we recall, has the highest proportion of academically based respondents, and
the political culture of academia with respect to nuclear weapons has likely been carried
over to that center. In any case, there clearly is not consensus among the British and
French scientists with respect to their countries' nuclear policies, and there clearly
is consensus among the Germans and Italians with respect to theirs.

There is a certain amount of deeply rooted emotional opposition to nuclear weapons

Table 12. Approval of Nuclear Weapons for Own Country, by Laboratory and Nationality (%)

	ESTEC				CERN				ISPRA		
---	Br	Fr	Ge	It	Br	Fr	Ge	It	Fr	Ge	It
Yes	57	58	5	0	40	38	6	11	52	6	13
No	29	26	80	100	52	62	88	79	44	94	74
DK	14	16	15	0	8	0	6	11	4	0	13

among scientists at all the laboratories, and many of the British and French who did favor nuclear possession by their own country said that they did so reluctantly or noted (particularly the French) that they had not favored the initial decision to build a nuclear force, but now that the government has spent the money, there was no sense in simply throwing away the results. Overall, however, opposition to nuclear weapons was not as strong as has been expected. Although acceptance of national nuclear policy by scientists of those nations that already possess nuclear weapons was far from universal, and weakest in the most basic research-oriented institution (CERN), there was substantial support for national nuclear weapons.

With regard to possession of nuclear weapons by "Europe" rather than any single European nation, the picture is quite different. Proportions favoring possession among the British and French contingents in all laboratories are reinforced, the Germans do a complete about-face and show strong majorities in favor, and the Italians also follow suit, although not as strongly. Table 13 shows the distribution for this question, which read literally, "Would you favor possession of such weapons by a future European military force?"

It is apparent that, while healthy minorities of three nationalities at CERN still oppose nuclear weapons, much of the opposition to national nuclear weapons does not carry over to a European force. We may take this as an indication that, for many of the scientists, opposition to national nuclear weapons is based less on moral grounds and more on grounds of scale, cost, and utility, all of which make national possession a less attractive proposition for individual European states.

Although there is recognition that possession of nuclear weapons would be inevitable and probably necessary in a unified Europe, the scientists' approval of the concept tends to lack enthusiasm. There is a feeling that an integrated Europe would probably act more responsibly than any single nation in handling such a force, and that its possession is part of regaining the status of "a leading continent." A few quotations from the interviews might suggest the tone of this response:

Table 13. Approval of Nuclear Weapons for Europe, by Laboratory and
Nationality (%)

	ESTEC				CERN				ISPRA		
	Br	Fr	Ge	It	Br	Fr	Ge	It	Fr	Ge	It
Yes	60	58	75	43	52	62	53	37	60	71	49
No	29	11	10	57	40	38	24	47	16	17	31
DK	12	32	15	0	8	0	24	16	24	11	20

Well, I told you I am very reluctant to consider these questions. Basically, I would
say nuclear weapons as such are very undesirable. As there are certain nations who
have nuclear weapons in their possession, however, we have just to face the fact and
in this respect I think that, if we have a European political community and a European
force, this force would have to be equipped with nuclear weapons because otherwise...
well, the politicians wouldn't be in a position to argue with other politicians who have
nuclear weapons.
—German ESTEC scientist

If there was really a European force, as I was telling you previously, an army is
meaningful only if it is powerful at an international level and, as for being powerful
at an international level, well, you need nuclear weapons.
—Italian engineer at CERN

I would like to see all nuclear weapons destroyed. In the event they are not, and there
should be nuclear weapons, I think that a European force should have them—but not
any individual country.
—French engineer at ISPRA

This is another aspect of the independent role for Europe which was discussed earlier.
Reluctant acceptance of nuclear weaponry is a price that must be paid for genuine political independence.

Generally, questions concerning nuclear weapons were treated more articulately and
in greater depth than other aspects of East-West relations and military security. The
scientists, in all laboratories and across all the nationalities, seemed to have devoted
more thought to such matters and had better-structured ideas about them. Nevertheless,
the interviewer could often sense a certain coldness that developed when this area was
brought up. The respondents seemed almost relieved when the interviewer went on to
the next subject area. If there is an area of political and international life which the
scientists would prefer not to think about, this is evidently it.

INTERNATIONALISM VERSUS REGIONALISM
A preference for supranational European forms over individual nation-states is evident
beyond question in the data that have already been reported. In their strong consensual

approval of transnational activities in the economic, political, and military domains, the respondents testified to their recognition of the growing interdependence among the various Western European countries. Aware, however, of nationalist trends, even at the subnational level (as visible during 1967 in Brittany, Wales, Scotland, and Flemish-speaking Belgium, for example) and desirous of preserving the unique character of Europe which is based on the cultural individuality of its parts, they envision a federated formation for the region, maintaining common policies toward external affairs, developing the economies of economic and technological scale, and yet retaining considerable local autonomy.

In opting for transnational solutions to the major political and social problems of the age, the scientists do not differ qualitatively from other influential classes in European society. To quote once again from the conclusions of Lerner and Gorden's elite panel survey:

Indeed, there has been a convergent consensus in Europe, over the last decade, that national options are not viable and that transnational choices are the only realistic alternatives. We have witnessed the passing of nationalism in the form familiar to previous generations and even to the early years of the generations now in charge. [60]

It is not any wild idealism that has shaped this consensus; it is merely pragmatism —a realistic appraisal of the desirable and the possible. The scientists differ from the nonscientific elites mainly in the intensity of their desire for closer integration and the breadth of their consensus on the shape of Europe, which transcends, for the most part, differences in national viewpoint. A "Europeanist," that is, regionalist, outlook permeates their political thought.

We have already seen, however, that with regard to supranational military integration at least, a certain degree of interest in the United Nations (and hence extra-European internationalism) exists (Table 11). Indeed it is among the British—that nationality whose attachment to the visions of European integration was relatively weaker than the others—that interest in the United Nations is strongest. One is tempted to inquire, on the basis of this finding, about the degree to which European regionalism and wider internationalism represent conflicting loyalties, or at least competing scopes of interest. Two questions dealing with extension of the United Nations into a world government were asked of the respondents, and analysis of their numerical tabulation enriched by consideration of the interview responses provides some insights into this matter.

The first question demanded an evaluation of the prospects for international organization on a worldwide scale. Unfortunately, however, it related the general problem to the fate of a particular body, the United Nations, whose fortunes were at a relatively low ebb at the time of the survey. The question read: "Do you think that the United Nations can be transformed eventually into a world government?" Expectations were not very strong. Less than one-fourth of the total number of respondents (24%) thought that it could. Fourteen percent rendered no judgment, while almost two-thirds responded negatively. Part of this pessimism reflected a low evaluation of the United Nations as a viable organization rather than skepticism of the world government idea. A large number of write-in comments (35) to this effect testified to the strength of this feeling.

On the other hand, despite their low expectation, the willingness of the scientists to approve such a world union was high. Responding to the question "In principle, would you favor your own national government giving up a certain amount of its sovereignty to participate in some form of world government?" nearly three-quarters (74%) of the scientists said "yes", while only 15% opposed the idea.

Of course, these two questions measure only a very limited aspect of what has been called here "wider internationalism." A real understanding of the concept of "internationalism" would require a much broader range of questioning than it was possible to employ in this study. For this reason, we have chosen to deal with the elements of receptiveness to, and expectation of, a political community roughly analogous to a European community but on a worldwide scale. Approval of such a community evidently implies formal surrender of a major part of the power of national self-determination to a body in which the majority of nations do not share either the culture, the values, or the historical traditions of Europe. This, one might argue, is as good a heuristic indication of internationalism as any. The general willingness of the respondents to see their nation surrender sovereignty to such a hypothetical community is a profound statement indeed. Despite the fact that we have no other elite (nonscientist) data with which to compare these responses, we cannot help but be impressed by the strong vote of confidence given the notion of worldwide political community by these scientists. The impressiveness of this response is qualified only by the fact that such a world government is so obviously an ideal under present conditions that constraints of reality, which might otherwise shape response patterns into more conventionally nationalistic forms, are relaxed. Internationalism might be relatively easier to express at this level of abstraction than on more mundane issues.

In any case, it is necessary to ask what, if any, relation do these patterns have to the patterns of approval and expectation with regard to European integration? First of all, for most respondents there appears to be little interaction between receptiveness to European unification and approval of world government. Among these respondents who approve either union of the Six or a larger type of European political

integration or both, 75% would approve surrender of sovereignty to a world government. Among those who oppose any type of political union in Europe, the proportion
favoring world government was nearly the same (68%). This is one good indication
that European regionalism and worldwide internationalism do not conflict in the
minds of the respondents. Further, it suggests that most of the small minority of
respondents who did not display strong regionalist sentiment were nevertheless not
narrowly nationalist in their outlooks. The absence of interaction was found within
the ESTEC British (where the greatest number of non-Europeanists came from) to
the same degree as in the sample as a whole. Examination of other questions concerning acceptance of European integration yields parallel results.

On the other hand, expectations of European integration, not of the Six, but of
"Big Europe," do seem to be linked with expectations of world government. Table
14, from which these respondents who were not asked about larger European integration have been eliminated, shows the cross-tabulation of the two questions (in
raw numbers, not percentages).

The hope that the United Nations might someday be transformed into a world government is much stronger among those who believe a wider European union is feasible
(35 out of 96) than among those who do not (26 out of 171). This linkage appears as
well within those laboratory/nationality groups that are most skeptical of world government. Otherwise, within and across these groups, no evident differences (such
as by seniority, professional level, and so on) appear between those who seem optimistic with regard to the likelihood of world government, and those who do not.

Once again, though, the division is perhaps not as deep as the tabulations make it
seem. None of the respondents (judging from interview responses and comments)
envisioned a world government with anywhere near the same degree of concreteness
as a united Europe, while at the other extreme only a few regarded it as pure wishful
thinking, without any possible basis in reality. Most respondents viewed it as a distant

Table 14. Expectation of Wider European Union by Expectation of World
Government (n = 307)

		Expect World Government		
		Yes	No	DK
Expect European Union	Yes	33	55	8
	No	26	125	20
	DK	8	14	18

goal, something for which to hope but not expect to see very soon. Thus, while the scientists who were most optimistic about wider European union were also most optimistic about world government, there was no competition in preferences, no need to choose between Europe and the world. Further, there was no evidence to indicate that growing enthusiasm for European integration might affect a scientist's judgments on possibilities for wider unity.

While a very few respondents mentioned that they thought European integration might have detrimental effects on the Third World—if the European countries did not take steps to assure free trade with the developing nations—most saw no conflict between the building of Europe and the development of greater world unity. The most common view expressed in the interviews, in fact, reflected the idea that an integrated European state, rather than the individual European countries, could in some remote future become part of a broader world federation. Thus, the question asking whether the respondent would agree to his country giving up some of its sovereignty to participate in a world government was often answered in the following manner:

As I said to you, I am already for giving up German sovereignty for a European government. Now if this process could be continued, I certainly would like to have finally a united world and everyone a citizen of this ... it would really be the ideal. But I do not believe it will happen very soon.
— ISPRA German scientist

Well, I wouldn't put it like that. I would say that by the time we get to a world government, you're not talking about a British authority or a French authority, you're talking about a European government. And it's the European government that would be giving up some of its sovereignty to a world authority.
—ESTEC British engineer

This is, I think, the logical development of uniting Europe politically ... It's just an analogy between the single countries first, then regions, then finally the logical consequence is a political world government.
—ESTEC German scientist

Regionalism is recognized as the more realistic prospect, but not at the expense of broader internationalism.

The same sort of reasoning that led respondents to the conclusion that integration of the European states is a necessity was extended to a worldwide scope. In the words of one Dutch EURATOM scientist: "I don't believe in a real independence of countries; I think this is impossible." Nevertheless, the problems confronting broader unification are formidable indeed, and respondents' ideas of such unification were necessarily quite vague. This same Dutch EURATOM scientist provides a typical example:

I can imagine that a world government could exist. I don't know exactly how it would be made and how it would be controlled. At the moment we think that democracy is the best way of government, but if this is also possible with the entire population of the world I don't know. Personally, I hope we find some way to do it.

Of all the subject areas on the political portion of the questionnaire, this one seemed to evoke the least emotional involvement. The international scientists, without losing the basic idealism of the scientific approach, are too sophisticated to place much stock in this notion. The words of an ESTEC Briton provide an appropriate summary of our discussion:

I would certainly favor a world government, but I think to work for a united Europe is a far more promising prospect at the moment. That's the only place I would put any real effort into now.

INTERNATIONAL LABORATORIES AND EUROPEAN INTEGRATION: CONCLUSIONS AND REFLECTIONS

This study did not set out to provide definitive answers to the wide range of questions that it opened up. It was in essence a reconnaissance, a first attempt at examining some highly interesting but unexplored territory. Thus, our conclusions and reflections are presented here in a cautious and rather tentative voice and should be treated accordingly.

One finding does emerge most clearly from the data on the scientists' political attitudes—that an attitude pattern highly favorable to increased political integration in Europe is practically universal among scientists of all European nationalities in all of the laboratories surveyed. Because of the strength of this consensus, however, it proved all but impossible to detect variations in attitude structure which might be attributed to effects of the laboratory experience.

The scope of agreement on most of the major European issues was strong enough to overwhelm any internal divisions. Support for the existing organs of European collaboration was so universal that it could easily be, and in fact was, taken for granted. Beyond this, the data revealed nearly unanimous approval for political integration of "little Europe" and expansion of the core group of six nations to include other Western European powers. Expectations on these matters did vary, although not in any regular fashion. Motivations were traced to a common hope, that of seeing Europe regain its world position as a "leading continent." The availability of reported data and conclusions from other elite studies permitted certain comparisons to be made. Ideally, it would have been desirable to base such comparisons either on scientists in national establishments or on nonscientists in other international organizations, so as to control independently for the scientific influence and the international influence.[61]

Failing this, comparisons with national elite panels did at least demonstrate that the scientists in this study displayed what seemed to be a more intense desire for close and rapid political integration, and certainly a much more broadly transnational consensus on these European issues than the national elite panels.

Differences between the various laboratories and national groups turned up primarily on issues where national situations diverged widely (such as national nuclear weapons), or where the character of a particular laboratory suggested a strong influence in one direction or another. On matters concerning European integration, the most consistent deviators were found among the British. While most of the British scientists presented attitude patterns that paralleled their colleagues from other European nations (and thus differed significantly from their nonscientist compatriots), significant numbers of British scientists and engineers, particularly at ESTEC, reflected their domestic political climate in showing a distinctly lower degree of European sentiment. It was in this group that increased experience in the international laboratory seemed to make a difference: the proportion of Europeanists was substantially higher among those ESTEC British with high seniority than among those who were relatively new in the international laboratory.

All of this points to the conclusion, heavily qualified as just described, that among most of the scientists there is not a great deal of room for movement with regard to European orientation. The vast majority were strongly in favor of the European idea before they came to the international laboratory and thus did not need to be "converted" by the experience. Among the various political issues explored in the survey, European integration is clearly the most salient to the scientists.[62] The international laboratories, however, serve mainly to bring together scientists with similar views on this matter, rather than take scientists with diverse views and shape them into a consensus. The invariance of this European sentiment was particularly striking in view of the variety of motives that brought the scientists to the laboratories, as well as the profound variations in organizational morale. Apparently, the stronger professional orientation of CERN scientists relative to their opposite numbers at ESTEC and ISPRA does not imply a lesser degree of Europeanist sentiment at CERN. Furthermore, the differing experiences which ISPRA, ESTEC, and CERN have had with the politics of European integration and their differing organizational contexts—mirrored so clearly in the morale at each—have not produced measurable effects on the scientists' political views. Despite the fact that ISPRA scientists in particular often pointed discouragingly to the frustrations of trying to "build Europe" and to the irrational behavior of the Six partners, their preferences and expectations were remarkably similar to those of scientists at ESTEC and CERN.

Although our data do not show the laboratory as an attitude-shaping experience for the international scientists, they do reveal some intriguing things. In itself, the existence among the scientists of such a broad consensus on European issues is an important finding. Even if the scientists' attitudes do not change measurably through the laboratory experience, it is likely that a significant reinforcement effect does occur. Thus, while such an indication cannot be seen directly in the data, the scientists' commitment to the concept of European integration may be deepened, or at least maintained, by this aspect of the laboratory environment. Beyond this, the importance of bringing together a large number of individual scientists with similar ideas might lie in the opportunities thus created for mobilizing the latent political force that exists within this group. Again, the data cannot support this idea directly, but in revealing the strong consensus, they at least allow for such possibilities as shaping the scientists into an associational interest group aimed at European integration.

Interest groups do not spring up by themselves, however, and we must also consider the matter of relating attitudes and behavior. The data permit us to speculate on the ways in which the strong Europeanist attitudes held by the scientists might be translated into action. In exploring the general framework of attitudes, we discussed the absence of a deep emotional involvement in most political issues. We found that concern with political matters tended to be placed on an intellectualized, detached level by most of the scientists. Activities of an interest- or pressure-group variety would not appear as strong possibilities under these circumstances.

The somewhat delicate situation of the international laboratory environment does not favor any form of direct political participation, as we noted earlier. For the scientists to join in partisan or national political affairs would in most cases be a violation of their international civil servant status. Even voting, if not legally excluded, is at least a bothersome chore. Under these conditions, it seems appropriate to look beyond the immediate, somewhat insular setting of the international laboratory and examine briefly communication and career patterns of the scientists. To whom might they talk? Where might they go where their influence could be felt?

Several points bear mention here. CERN, ISPRA, and ESTEC, it will be recalled, had quite different policies toward personnel interchange, with ISPRA offering primarily tenured appointments and CERN and ESTEC encouraging limited terms of employment. (In the case of CERN, however, a large number of scientists had managed to achieve indefinite terms.) The data suggest that the longer a scientist remains in an international laboratory, the less likely he is to want to return to his home country.

In response to a question on future plans, fewer than half of the respondents indicated a desire to return to their own countries, at least in the foreseeable future.
Many of the scientists were anxious to remain in an international environment of one
type or another, still others hoped to come to the United States. [63] Overall, a significant proportion of the scientists are likely to remain in situations where the
probability of their becoming active politically (in European affairs), in a conventional
sense, is low. The remainder, however, through returning to institutions in their
home countries, will at least be able to take part in domestic political activities.
Furthermore, in their day-to-day collegial relationships they have the potential of
helping to shape the political attitudes of a much wider set of scientists, and, to the
extent that they maintain professional ties created through the international experience,
they may effectively internationalize the European scientific community. [64]

The notion of the scientists returning to institutions in their own countries brings
us back, approximately, to where we started out. The government sector is of particular interest. Ostensibly "nonpolitical" participation in government, either through
roles in administration or scientific advising, could be an important avenue through
which the Europeanist attitudes of the international scientists can exert an influence
on the process of European integration.

About one-third of all those respondents who did not come directly from their
studies came from government posts. While many had been engaged in research, a
significant share, primarily British and French, could be classed as middle-level
technical administrators. Most of these were on temporary leave (secondment) and
planned to return to their former posts. They, plus those scientists who did not come
out of government but will enter its service upon leaving the international laboratory,
comprise a major proportion of the scientists with a political action potential.

The real influence that the scientists' political views may exert in such roles is
difficult to assess. Three limitations on their political influence are, however, worthy
of mention. First, the scientist's nonpolitical interpretation of his role probably decreases his propensity to apply his own political orientations to the problems which
he encounters. In addition, the fact that most of the government posts which scientists would be likely to occupy are in technical agencies, often in the administration
of research and development, no doubt limits the scope over which their political influence may be exercised. Finally, international experience is not always positively
valued by national administrations. While British respondents often seemed to believe
that their domestic status would be enhanced through their participation in an international venture, many Frenchmen reported that their colleagues in government ser-

vice at home looked down upon those who went abroad to work, particularly in an international organization.

Although the "scientific advisory culture" is not as widely diffused among European governments as it is in the United States, the role of scientific advisor probably has the widest potential appeal of all the various forms of political participation open to the international scientists. This appeal is a function, in large measure, of the fact that becoming an advisor does not entail conflicting role obligations. Advisory work is viewed as technical and nonpolitical, allowing the scientist to retain his virginal, apolitical status as well as his professional commitment to research. Because of these characteristics, it is also a form of activity through which a scientist may exert an influence in his own country while continuing to work in the international laboratory. Scientific advisory roles, however, are generally restricted to a rather small number of individuals, particularly those with strong scientific reputations that were established before they went to the international laboratory. The scientist's ability to develop his domestic reputation, once at the international center, is limited (sometimes enhanced, of course) by the laboratory's scientific standing as well as by the network of personal contacts he is able to maintain.

Through the role of advisor, the scientist is often dealing with broadly based and important issues in which science is a vital part of the context but whose effects are felt by large areas of society. In the European setting, the presence of such roles of men whose basic political inclination is toward European integration is certain to assist that cause. It is recognized that, in providing what is ostensibly technical advice, the scientist is incorporating his own underlying political assumptions. This theme has been much discussed, most often from the point of view of the policy maker whose job it is to separate the technical content of the advice from its underlying political tone. It would be naïve to expect dramatic results from this sort of influence in a short period, but as the science advisory culture grows in Europe, and as international scientists take their places within it, its effects are likely to be felt.

NOTES

1. For her invaluable assistance in carrying out the study upon which this chapter is based, the author wishes to thank Carolyn R. Teich.

2. For an analysis of this belief, see Jean-Jacques Salomon, "The Internationale of of Science," Science Studies, Vol. I, 1971, pp. 23-42.

3. "The Vienna Declaration," Statement from the Third Pugwash Conference, held at Kitzbühel and Vienna, Austria, September 14-20, 1958.

4. Ernst B. Haas, <u>Beyond the Nation State</u> (Stanford, Calif.: Stanford University Press, 1964). See especially pp. 21-22, 47-50.

5. The two best analyses are Lawrence Scheinman, "Euratom, Nuclear Integration in Europe," <u>International Conciliation</u>, No. 563, May 1967, and Stanley Hoffman, "Obstinate or Obsolete? The Fate of the Nation-State and the Case of Western Europe," <u>Daedalus</u>, Vol. 95, Summer 1966.

6. The survey was carried out by the author during the period from April through July 1967. Five smaller laboratories, incorporating sixty-four additional respondents, were also included in the study but are not discussed here. Full details on the methodology may be found in the dissertation from which this chapter is condensed, Albert H. Teich, <u>International Politics and International Science: A Study of Scientists' Attitudes</u>, Ph.D. dissertation, Department of Political Science, M.I.T., 1969.

7. It should be noted that a number of unavoidable factors made the sample somewhat asymmetrical; for example, since Britain was not a member of EURATOM at the time of the study, there were virtually no Britons at ISPRA.

8. The acronym "CERN" is derived from the original French name of the organization; Conseil Européen pour la Recherche Nucléaire.

9. The members at present are Austria, Belgium, Germany, Denmark, France, Great Britain, Greece, Italy, the Netherlands, Norway, Sweden, and Switzerland. Spain was a member at the time of the survey but has since dropped out.

10. Lew Kowarski, one of CERN's founders, has written an authoritative history of the organization, "An Account of the Origin and Beginnings of CERN" (Geneva: CERN Document, 1961). For a rather less official version of the same story, see Robert Jungk, <u>The Big Machine</u> (New York: Scribner's Sons, 1968).

11. Since the completion of this study, there have been some difficulties among the member states concerning the construction of CERN II, a new, larger accelerator. While these were eventually resolved, it is likely that the uncertainty about CERN's future did have some temporary negative effects on morale.

12. The most important of these studies was "Some Aspects of the European Energy Problem," prepared by Louis Armand and published by the Organization for European Economic Cooperation (OEEC) in Paris in 1955.

13. EURATOM's story is a complex one that can be only touched upon here. The dissertation upon which this chapter is based goes into considerably more detail. The interested reader is also referred to Richard Mayne, <u>The Community of Europe</u> (London: Victor Gollancz, 1962); Lawrence Scheinman, "Euratom: Nuclear Integration in Europe," <u>International Conciliation</u>, No. 563, May 1967; Jaroslav Polach, <u>Euratom</u> (Dobbs Ferry, N.Y.: Oceana Publications, 1964), and H. L. Nieburg, "EURATOM: A Study in Coalition Politics," <u>World Politics</u>, Vol. 15, July 1963, pp. 597-622.

14. EURATOM's budget figures vary from one source to another, and this figure represents the author's best estimate based on a variety of sources. It excludes separate allocations for a special reactor project (ORGEL).

15. The low personnel turnover is a result of the staff's <u>fonctionnaire</u> status and the fact that most of ISPRA's recruitment was done in a space of about two years. The

rapid buildup and the ensuing financial restrictions meant that there were few new places available after 1962-1963, and the <u>fonctionnaire</u> status meant that few people left.

16. COSPAR is the Committee on Space Research of the International Council of Scientific Unions.

17. The members of ESRO in 1967 were Belgium, Denmark, France, Germany, Italy, the Netherlands, Spain, Sweden, Switzerland, and the United Kingdom. In December 1972, representatives of the member nations of ESRO and ELDO (the European Launcher Development Organization) agreed to merge the two organizations into a new European Space Agency (ESA). Whether this merger would actually be consummated and what the role of the new organization might be remained unclear at press time.

18. <u>European Space Research Organization General Report 1964-1965</u> (Paris: 1966), p. 109.

19. Official figures as of June 1, 1967. The total does not include about 30 persons working for the ESRO Control Center, then at Noordwijk, but since moved to Darmstadt, Germany. The staff has since expanded.

20. In mid-1967, the ESTEC site had not yet been landscaped. It is doubtless far more livable now than it was then.

21. A statement of the director-general, Professor H. Bondi, following the successful launching of ESRO-2, testifies to this morale boost: "In every major endeavor there is a somewhat anxious and frustrating period between the decision to start work and the gathering of first fruit. . . . When at last IRIS (ESRO-2) was in orbit and operating, ESRO could begin to wear a different and far more self-confident face. Our purpose is to satisfy our customers, the space scientists of Europe. To see them gathering satellite data for the first time made us all feel proud of our work and brought home to us forcefully the whole aim of our activities." <u>ESRO/ELDO Bulletin</u> (Paris), No. 2, August 1968, p. 4.

22. Although questionnaire subjects were given their choice of English, French, and German forms, and 21% took French, while 14% took German, the French and German forms were more a matter of convenience than necessity for the respondents. It is evident that the study could have proceeded substantially unhindered if it had been done entirely in English.

23. See the accounts in Robert Jungk, <u>Brighter Than A Thousand Suns</u> (New York: Harcourt, Brace and World, 1958), especially Chapter 3; and J. D. Bernal, <u>The Social Function of Science</u> (Cambridge: M.I.T. Press, paperback edition, 1967), pp. 210 ff. Along a related line, see the chapter by R. D. Gillespie in this book, which discusses the politics of cybernetics in the Soviet Union.

24. These traits seem to be distilled from the more common general stereotypes, such as those of Frenchmen and Germans systematized by Erich Reitgrotski and Nels Anderson, "National Stereotypes and Foreign Contact," <u>Public Opinion Quarterly,</u> Vol. 23, Winter 1959-1960, pp. 515-528.

25. By professionalism we are referring to a qualitiative judgment of the individual's commitment to his subject, something discussed earlier in regard to reasons for coming to the laboratory.

26. Even the commission has taken official, if belated, note of this situation, pointing out in its 1967 annual report that political problems have created an atmosphere in the Joint Research Centers "which is unpropitious to the normal pursuit of the work and which must not be allowed to persist if the morale of the researchers is to remain unimpaired." Commission of the European Communities, First General Report on the Activities of the Communities, 1967 (Brussels-Luxembourg: 1968), p. 297.

27. When it is necessary to make long-term choices at CERN, such as those involving a child's education, they are likely to have an impact on the scientist's career. According to the CERN Courier, August 1968, "the education factor is a limiting factor in the recruitment of staff."

28. Parts of this section are adapted from Daniel Lerner and Albert H. Teich, "Internationalism and World Politics Among CERN Scientists," Bulletin of the Atomic Scientists, Vol. 26, February 1970, pp. 4-10, which reports on an earlier study of CERN.

29. Our finding here confirms that of Donald A. Strickland, who writes that scientists "often mean by 'political' that an activity is controversial, or that it has a high emotional content (is 'irrational' behavior) ...." Scientists in Politics (Lafayette, Ind.: Purdue University Press, 1968), p. 94. In addition, the senses of the words politique in French and politisch in German are slightly different from the English "politics," leading to some confusion.

30. The remainder of the distribution was "rarely" 18%, "occasionally" 45%, "often" 23%, and "very often" 11%.

31. Several psychological studies of scientists such as Anne Roe, The Making of a Scientist (New York: Dodd, Mead and Company, 1952), p. 58, and Bernice Eiduson, Scientists: Their Psychological World (New York: Basic Books, 1962), pp. 89, 94, 95, probe the depth of professional involvement and its effects on other activities.

32. One scientist who has made a deep emotional commitment, the Russian academician Andrei Sakharov, still demonstrates the desire to extrapolate science in his widely publicized manuscript "Thoughts on Progress, Peaceful Coexistence and Intellectual Freedom," New York Times, July 22, 1968. A method is "'scientific'" he states, if it is "based on deep analysis of facts, theories and views, presupposing unprejudiced, unfearing open discussion and conclusions." His proposal for cooperation rests on the notion that "international affairs must be completely permeated with scientific methodology," and further that "scientific methods and principles of international policy will have to be worked out."

33. The notion of incorporating an information test in the interviews and questionnaires was rejected as being likely to antagonize the subjects.

34. See Lerner and Teich, "Internationalism," p. 6.

35. One of the most clear-cut demonstrations of this is given in Bernard R. Berelson, Paul F. Lazarsfeld, and William N. McPhee, Voting: A Study of Opinion Formation in a Presidential Campaign (Chicago: University of Chicago Press, 1954), pp. 26-28.

36. The empirical literature in this area generally deals with "leftism" and "rightism" as expressed in party preference rather than on an independently generated scale, further complicating our problem. See Seymour M. Lipset, Political Man: The Social Bases of Politics (New York: Doubleday & Co., 1959), as well as Edward C. Dreyer

and Walter A. Rosenbaum, eds., Political Opinion and Electoral Behavior: Essays and Studies (Belmont, Calif.: Wadsworth Publishing Co., 1966).

37. The next few paragraphs are adapted from Lerner and Teich, "Internationalism," p. 6.

38. Daniel Lerner and Morton Gorden, Euratlantica: Changing Perspectives of the European Elites (Cambridge: M.I.T. Press, 1969), p. 127.

39. The author's earlier analysis of data from CERN alone, Lerner and Teich, "Internationalism," reached this same conclusion.

40. Similarly Lerner and Gorden, Euratlantica, p. 132, report that their national elite panels "had not yet acheived a consensus among themselves as to the next steps for the European Community, nor even to the direction these steps might take."

41. Karl W. Deutsch, Lewis J. Edinger, Roy C. Macridis, and Richard L. Merritt, France, Germany and the Western Alliance (New York: Charles Scribner's Sons, 1967), p. 74 and pp. 160-163.

42. Lerner and Gorden, Euratlantica, p. 144.

43. It is worth noting parenthetically that some respondents interpreted the question of wider political union as including the Eastern European nations and responded to this. So broad a union was not intended in the question, but its consideration as a possibility is certainly significant.

44. Including the "yes" as well as the "no, but later" response.

45. Lower interest, less discussion, and "rightness" are related to each other in the whole sample; see the previous section.

46. One might hypothesize on the other hand that the non-Europeans tend to leave before they acquire high seniority rather than changing their political views. Mild support is lent to this notion by the fact that the low-seniority deviates report a substantially lower level of job satisfaction than low-seniority consensuals.

47. The most popular write-in comment on this question was "after de Gaulle."

48. This finding, at least with respect to German elites, parallels that of Deutsch, et al., Western Alliance, p. 165.

49. Deutsch, et al., Western Alliance, p. 77.

50. Ibid., p. 164.

51. Lerner and Gorden, Euratlantica, p. 74.

52. While we do not have quantitative data that might directly support this notion, the interpretation is based on qualitative impressions from the interviews. Few of the respondents appeared to have any real liking for the economic or political systems of the countries of Eastern Europe, but the "Europeanness" of this region, based on cultural and historical ties, is highly valued by nearly all. The logic is expressed well by a British scientist who spoke of a "viable economic unit and sensible political unit." Relations with Eastern Europe are also treated in the following section.

53. It seems likely that this question would have gotten more useful results had it asked directly about supranational versus international government, which is the real heart of the issue.

54. See along these lines, Jean-Jacques Servan-Schreiber, The American Challenge (New York: Atheneum, 1968), and Lerner and Gorden, Euratlantica, Chapter 1.

55. Deutsch et al. Western Alliance, make it clear that this is not the case in their broader German elite sample, pp. 170-171 and p. 153, but note that there is "widespread sympathy" for greater contact with Eastern Europe among French elites, p. 66.

56. It is noteworthy that the percentage favoring this notion among leftists (66%) is not far different from that among rightists (53%).

57. Lerner and Gorden, Euratlantica, p. 106.

58. Deutsch et al., Western Alliance, p. 285.

59. Daniel Lerner, "Interviewing Frenchmen," American Journal of Sociology, Vol. 62, September 1956, pp. 187-194.

60. Lerner and Gorden, Euratlantica, p. 241.

61. Not knowing how the scientists in these laboratories compare with other European scientists no doubt places major limitations on the utility of these data. Resource constraints did not permit us to include this control, and useful hard data from other surveys were not found.

62. Of course the range of questioning was not extremely wide, and the respondents' perception of the purpose of the study may have influenced this finding.

63. Given the changes in American science since 1967, it is likely that the perspective on this has changed.

64. European professional societies (for example, the European Physical Society) are already beginning to emerge. See William H. Jahsman, "Toward a European Scientific Power," Technology Review, Vol. 73, June 1971, pp. 10-13. A study of the way that the social structure of European science has been affected by the international laboratories would be an important contribution.

Daniel Lerner

The doctoral dissertation by the late R. David Gillespie, from which the following paper has been extracted, makes an important contribution to the annals of science and public policy. Throughout history, the relations between men of knowledge and men of power have shaped the contours of society. In the twentieth century, the efflorescence of science—particularly of the highly visible technologies produced by applied science—has enlarged the scope of policy concern with new knowledge to include several domains of social life. At the same time, the political system of modern totalitarianism has sought ways to enhance its own power by the new knowledge without enhancing the authority of the men who produced this knowledge.

There have been similar conflicts between men of power and men of knowledge in the past. Galileo and the Inquisition are the familiar symbols of those conflicts. They have always produced deformities of knowledge and grotesqueries of power that were sometimes transient, sometimes permanently disabling. One may well ask if Spain would have fallen so far behind the other great nations of Europe over the past four centuries, despite the defeat of the Spanish Armada, had the Inquisition been less effective in muzzling new knowledge.

The modern totalitarianisms of Europe—fascism, falangism, nazism—were short-lived. Their efforts to harness the technologies of destruction to warmaking were defeated by the superior technologies (quantitative and/or qualitative) of their adversaries. Their putative long-term impact upon science must therefore remain a moot question.

The most durable of the modern totalitarian states is, of course, the USSR. In its half-century of kaleidoscopic history, the Soviet state has lived through a diversity of experiences in the continuing contestation between men of knowledge and men of power. The men of power have never really "let go" of their reins. But, on a number of occasions, often despite their systemic propensities, the Soviet men of power have been obliged to "let up" on their men of knowledge.

Such a relaxation occurred, perhaps most dramatically, when the Stalinist purges of the late 1930s left the USSR seriously weakened in top scientific-technological personnel, as it faced the Nazi challenge that brought it into World War II. Emerging from that bloodbath into the challenges of postwar "coexistence," the Soviet men of power then sought to reestablish their hegemony over all aspects of Soviet intellectual life. This was the "Zhdanov period," in the years preceding Stalin's death, when Marshall Zhukov was sent into eclipse, when the physicists Kapitsa and Landau were put under a cloud, when Lysenko was deputized as czar of genetics, when the DIAMAT (dialectical materialism) was reasserted as superior to all forms of "bourgeois science."

An outflow of this period—even after the "destalinization" of Soviet life under Khrushchev—was the continuing struggle over cybernetics. One interesting sidelight on this struggle was given me by the late Norbert Wiener, my colleague at M.I.T. and the inventor of cybernetics, who was the son of a professor of Slavic languages and an admirer of things Russian. At one point in the Soviet conflict over cybernetics—when both the science and its inventor were being reviled as creatures of bourgeois imperialism—Wiener told me that some of his own ideas were derived from the work of Soviet mathematicians, notably Kolmogarov. Moreover, Wiener added, despite official denunciatiamentos, Soviet work on cybernetics was continuing under different labels—an immensely important instance of "Aesopian language" in action to protect Soviet science.

Among Dr. Gillespie's contributions is the clarification of these Aesopian communications, particularly as the official policy shifted from dogged antagonism to euphoric embrace. On the zigzag courses and erratic consequences of this momentous shift, Dr. Gillespie sheds much light without heat. His workmanlike analysis merits careful reading.

It is the misfortune of all scholars in science and public policy as well as Soviet studies that this paper must remain Dr. Gillespie's major published work. His tragic death in 1972 truncated a promising career and deprived us of a respected friend and colleague.

R. David Gillespie

In the course of little more than a decade, the status of cybernetics in the Soviet
Union evolved from that of a hostile and dangerous "bourgeois pseudoscience" to one
of the principal means for the construction of the "material and technical bases of
communism." The development of the cybernetics dispute provides an excellent
means of gaining insight into the role and position of Soviet scientists in public affairs.
This study is an attempt to chronicle and analyze the dispute.

In the engineering fields of automation, remote control, and computer development,
one can see at least the public relationship between some scientists and the party in
pursuing new goals of communist construction. This involvment produced much comment from all sides on such major issues of Soviet public policy as the gap between
intellectual and physical labor, the differences between urban and rural life, the
role of education and leisure for the population, and the perennial problems of state
planning, labor productivity, and scarcity. To a large degree, the simplistic assertion of the party was that automation was bad for capitalism (being used merely to
exploit the worker) and good in socialism (being used to enhance his cultural and intellectual level and provide him with more leisure). From this there emerged a
clearer notion of science as a productive base in all societies (as opposed to part of
the "superstructure"). At the same time, the argument that the application of science
may have common features in all societies was steadfastly denied. [1]

More general problems of science and public affairs were encountered because
cybernetics in the USSR grew to embrace so many fields of science. In the Soviet
Union many of these fields, especially in the life and social sciences, have been
monopolized by "single official schools" under Stalin. The cybernetics discussions
provided one of the ways to seek the reestablishment of scientific multiformity in
these areas. Spokesmen for once-suppressed subschools began to reappear in the
post-Stalin system, sometimes riding the crest of the cybernetics wave. Their old
and new arguments were sometimes conducted from behind the twin shields of this
"new science" and their promises to contribute more to the solution of state and party
determined problems than had the monopolists. Khrushchev evidently chose to rely
upon the Soviet scientific community to a larger degree than Stalin, and the responses
from the community to his call for help were exceedingly diverse, embracing a very
rich division of scientific and social opinion.

This study is concerned with the process of scientific, social, and economic development in the Soviet Union. It would be methodologically pretentious to claim that

it presents a systematic interest group study of Soviet science and society. However, what may be observed are many scientific spokesmen, often caught between contradictory social goals or in the middle of jurisdictional disputes, attempting to combine their own interests, the interests of their own and other disciplines, and the interests of the state that they serve. While the Soviet setting provides some undeniably unique features and constraints, perhaps this overall view will not be entirely unfamiliar to other scientists working for other government sponsors.

BACKGROUND

Cybernetics was developed by Norbert Wiener and his colleagues in the West. The early work was carried out during World War II, when several American scientists and engineers directed their efforts toward obtaining more accurate systems for automatic gun control. The discipline was loosely defined by Wiener as the study of communication and information in self-regulating systems. The name of the new field, coined by Wiener in 1947, was derived from a Greek word meaning "helmsman." In 1948 Wiener published his "more or less technical book" (as he later described it) Cybernetics: or Control and Communication in the Animal and the Machine. This was followed in 1950 by his popularization, The Human Use of Human Beings, subtitled Cybernetics and Society.

Wiener and other proponents of this new branch of study enthusiastically described the power and generality of the basic concepts of cybernetics. Promises of wide potential for the new field were often based on the existence of control systems in various spheres of nature which were analogous to one another and which could therefore be approximately described by similar mathematical models. For example, a striking similarity was noted between the behavior of an affliction of the human nervous system (purpose tremor) and improperly designed gun control systems whose "positive feedback" caused the gun to oscillate more and more wildly rather than converging on the target. The early development of cybernetics in the West was in part pursued on the basis of such interdisciplinary analogues and embraced elements of information theory, game theory, physiology, mathematics, electrical engineering, and computer design.

With the advent of exciting new technical developments in these several fields, the imagination of many people in the West was captivated by the potential of cybernetics. This was partly due to the involvement of cybernetics with both the threats (for example, unemployment) and promises (for example, an easier life) of industrial automation. At a different level the question was posed: "Can machines think?" In response to these sorts of questions there occurred the proliferation of analogies, speculations, promises, and anthropomorphisms that produced very little practical benefit. At the same time, although the "technical discipline" of cybernetics never really progressed as such, pieces of the once-envisioned discipline were pursued and developed in a large variety of old and new fields of science. These included such areas as engineering control theory, mathematical control theory, computer design, operations research, game theory, information theory, and several others. As a result, one rarely (if ever) encounters a scientist or engineer in the United States who would characterize himself as a "cybernetician."[2]

In the Soviet Union the development of cybernetics followed a very different path. There the alleged pretensions of universal applicability for cybernetics were attacked at the time of Stalin's death in 1953. The field was taken seriously by both its opponents and proponents in the following years, at least to the extent that its proponents and opponents seemed to posit it as a separate (albeit very vaguely defined) scientific discipline. Since cybernetics entered the argument as a somewhat amorphous discipline within which a large number of scientific pursuits could be integrated, the arguments tended to be conducted at two levels.

At the higher level, the overall discipline was attacked or supported on the basis of its necessity and/or legitimacy. The spokesmen occupied positions ranging from those who found nothing wrong with the then current integration of Soviet science to those who did not want independent study to be integrated either in the current system or beneath the rubric of cybernetics. Since the current system of integration at the most general level involved dialectical materialism and state control through science planning, these arguments usually were concerned with major issues of philosophy, ideology, and the politics of science.

At the more specific level, the Soviets showed a strong tendency to divert the argument from the more general concern of a major new integrative discipline and either to attack or support individually the diverse activities clustered within the proclaimed domain of cybernetics. Therefore, various spokesmen came to speak of such hybrid fields as "biocybernetics," "medical cybernetics," "economic cybernetics," and so forth.

By 1961 "cybernetics" had won the explicit approval of the Communist Party of the Soviet Union and found a place in the Party Program. Cybernetics quickly became "fashionable," and, since it remained very poorly defined, there existed a fine opportunity for the most diverse citizens to proclaim their own activities as belonging to the domain of cybernetics and its applications.[3] However, a more durable and interesting aspect of this development concerned the use of "cybernetics" as a shield behind which some scientists and schools of science reemerged in the Soviet Union after having been proclaimed anathema under Stalinism.

Owing to this diffusive and rather complex path of development, the present analysis is concerned both with the center of the Soviet cybernetics dispute and with the separate arguments that occurred at the periphery, in established fields of science. The development of independent scientific inquiry and technical innovation has often been marked as one of the more exciting aspects of Khrushchev's rule. Here it will be shown that in some areas technical and scientific renovation were often quite as controversial and important. Further, scientific innovation and renovation were often closely connected during this period of growth and relative independence in Soviet scientific life.

It is possible to separate two periods in the development of cybernetics in the USSR. During the first period, from 1953 to 1958, the public discussions dealt largely with the central problem of determining whether cybernetics should have a place as an independent scientific discipline in the Soviet scheme for science. The public discussions were often couched in terms of ideology and philosophy of science, and there was a lingering tendency to denounce cybernetics as one more "bourgeois pseudoscience in the service of capitalist imperialism." By 1958 this sort of simplistic argument no longer carried much weight because of Party demands to learn from the West and let "life and practice" be the arbiter of scientific correctness. In 1958, various institutions and publications concerned with cybernetics were established.

The years from 1958 to 1961 seem to have been a kind of "gestation period" for Soviet cybernetics, during which proponents for the field prepared books and articles advocating the use of cybernetics in the most diverse fields of intellectual work. In 1961, cybernetics won explicit approval in the Party Program. For several years after this recognition both the popular press and some more specialized journals were replete with exhortations to employ "the methods of cybernetics" in the most diverse fields of science and technology. Whereas the major topic of the first period was the legitimate existence of cybernetics itself, in the second period the questions addressed were more likely to occupy the periphery of the domain of cybernetics and be concerned with the propriety of using "cybernetics" in specific established disciplines.

It is likely that the use of cybernetics (often just as a symbol of mathematics and computers, or even of qualification itself) in such areas as biology, physiology, and sociology was the source of the passion in the arguments about ideology and philosophy which went on in the first period. Only later did these objections become explicit and apparent. During the earlier period there was a strong tendency to convert internal disputes within scientific fields into ideological and philosophical arguments of universal scope.[4] It is quite possible, in view of the apparently continual development of Soviet work on computer development and earlier work on automation and remote control, that the appearance of the early hostile articles marked the end of technical controversies which had been going on beneath the surface for some time. This may have been an "echo" of more purely technical disputes for which de facto technical resolutions had been found, that is, a final appeal to the nontechnical community.

The analysis will be concerned first with the central arguments about cybernetics which took place between 1953 and 1958. In order to understand the changes that were occurring in the Soviet system and their effects on Soviet science, it will be necessary to describe briefly the system of science as it was at the time of Stalin's death. That system, along with changes initiated in the years that followed, constitute the all-important context in which the debates took place. Within that context, we will consider the major arguments raised against cybernetics as an independent discipline, and the associated defenses of the science.

In the second portion of the analysis, the concern is with specific arguments that emerged later in various established scientific disciplines. These disputes were most often connected with the growth and development of these disciplines and the establishment (often, the reestablishment) of "subschools" within them. These arguments were often against dogmatism and "monopoly schools" in Soviet science, and new, independent approaches to scientific research were presented as concrete corrective measures in branches of science that had been lagging and/or constricted in their development during Stalin's rule.

By establishing ties between these sciences and the newly approved discipline, "cybernetics" often served as more a "Trojan horse" than "helmsman" in the reopening of debates about independent research in such moribund fields of Soviet science as economics, sociology, law, biology, physiology, and medicine. The appeal was to let Soviet scientists perform independently and according to canons of expertise developed in their own disciplines and then to "let life judge" the product of their labors. These appeals were often peripherally supported by the very practical contributions that once-controversial "directions" of science (for example, cybernetics and relativity theory) had made to Soviet military and industrial science and technology.

This introduction concludes with two warnings. First, the focus given here to the "political" uses of cybernetics should not be construed to imply that the Soviets have not made technical use of the science. It is clear that in some engineering and scientific fields, for example, computer design and control theory, they have made steady progress. The focus on the political uses of the science is simply a reflection of the fact that the contributions of cybernetics to some other fields have been at least as much political as technical in nature. The apparent expansion of cybernetics into these other fields has been seemingly unaccompanied by technical achievements of much impact either upon cybernetics or upon the penetrated discipline (for example, cybernetics and law). In other cases, cybernetics can be linked to another discipline only by the loosest use of the term. Thus, for example, considering econometrics as a part of cybernetics may have been appropriate in the Soviet Union at one time. Ever to have done so in the United States would have been at most a unique use of the words.

The second warning concerns the analysis of the second period in particular. It is the assertion of this analysis that intellectual ferment and drives for independent schools of thought led in some ways to the expansion of the domain of cybernetics in the USSR. It is not the assertion of this analysis that the expansion of cybernetics necessarily caused the ferment or even contributed much to the establishment of some of these independent schools. The latter assertion is usually unsupportable either because the issues at stake had taken form much before the advent of cyber- netics, or because similar discussions took place simultaneously in contexts far removed from "cybernetics."

With the attainment of a legitimate independent existence, some of the "subschools" that were at one time clustered behind the aegis of cybernetics may have recognized a positive interest in dissociating themselves from the science. Some have charac- terized this phenomenon as an impending Soviet "disenchantment" with cybernetics or the "end of the honeymoon" with the field. We may posit three types of reasons for this disengagement.

The first reason is that the old philosophical and ideological debates may not have been resolved except by the use of "administrative methods," that is, the pragmatic decision by the policy makers to use cybernetics for its utilitarian advantages. We still read of unnamed "reactionary" enemies of cybernetics who lurk in the corners of Soviet science and philosophy. Some of the disputes have a currency and freshness which indicate that little may have been really resolved, except for the establishment of independent subschools in some disciplines. For these groups, unessential links to cybernetics may be an unnecessary liability to their independence.

Second, many of the proponents of cybernetics in the service of various disciplines made extravagant claims that were often technically irresponsible, either because they were made in the passion of the dispute or because they were simply beyond the technical capabilities of the "state of the art."[5] In many areas it is likely that the con- crete, practical contributions of cybernetics to the Soviet system compare rather poorly with what was once promised. Given the unresolved disputes and unfulfilled promises, it is not impossible that controversies about cybernetics per se could break

out again, especially as the current regime continues a new stress on the applied
nature of scientific endeavor and the demand for concrete, economic payoffs from
research. However unlikely such an event may be, the new schools could minimize
the risk by breaking unessential ties with cybernetics and its sometimes irresponsible
advocates.

Finally, the use of cybernetics by the independent "directions" has long been associated with the difficulty of embracing cybernetics sufficiently closely to use its
power and appeal, but not so closely as to be subsumed entirely within some new
discipline. For example, during the development of bionics (the study of control
systems in living nature for application to technology) in the Soviet Union, one found
the same advocate of cybernetics (Academician Berg) once treating bionics as a part
of cybernetics, and later defining bionics as an independent discipline to which cybernetics and other sciences can make important contributions. It is clear that many of
the leading spokesmen of fields once on the periphery of cybernetics prefer to practice as economists, biologists, physiologists, and so on, rather than as "cyberneticians," computer experts, or control theorists. Cybernetics, once having "served"
some of theis disciplines with that which they wished to use (popular appeal and some
tactical strength), may have exhausted much of its political and technical usefulness
for them.

The foregoing trends are comparatively recent, and hence this portion of the analysis
is more speculative than the rest. If these speculations are correct, however, then it
is quite possible that cybernetics in the USSR has begun to follow a path similar to
that of its development in the USA: cybernetics may be losing its appeal as a discipline
and political symbol, but its technical contributions will continue to find important use
as one of many mathematical methods in the service of ther disciplines.

THE STALINIST SYSTEM OF SCIENCE

The publication of Wiener's <u>Cybernetics</u> in the United States occurred at the same
time as the campaign of cultural chauvinism in the Soviet Union. This was the period
of the so-called <u>Zhdanovshchina</u> (named for A. Zhdanov, who died in 1948), a campaign directed toward the development of a unique "socialist science"—lacing together
of elements of Soviet ideology and materialist philosophy with elements from the physical, life, and social sciences—that was counterposed with great intransigence to be
"bourgeois science" of the West. This construct was both insidious and powerful when
combined with the Soviet propensity to plan and support science along single lines of
development, and the period produced the dominance of single, officially supported and
presumably orthodox "schools of science" in several fields.

These schools were headed by officially recognized "orthodox interpreters" whose interests were to strengthen their monopoly positions with selected pronouncements of the masters (such as Pavlov in physiology and Michurin in biology) while attacking rival heads of "independent subschools" as bourgeois heretics. During the time of this intellectual constriction many leading figures of Soviet science were removed from positions of influence.

The Zhdanovshchina was ushered in with a frantic combing of texts to demonstrate Russian priorities in technological and scientific invention. This was the era when Russians discovered that Russians had discovered almost everything of enduring technological interest.[6] Coupled with this patriotic drive for technical and cultural superiority over the West was an extremely violent campaign to denigrate Western achievements. Cybernetics was by no means the first "bourgeois science" to be collected under the terms "pseudoscience," "misanthropic reaction," and so on.

Of greater moment than these vilifications, however, is the apparent fact that some scientists, ideologists, and party leaders took as their serious task the reorientation of Soviet science so as to bring forth two systems of world science, the one bourgeois and the other socialist. This was evidently to be done by a curious mixture of materialistic philosophy, ideology, state planning of science along unidirectional lines, and the various branches of science. Some of these intellectual conglomerates stuck together better than others, and indeed some of them seemed to constitute viable, mutually supportive, and officially supported "schools" of Soviet science well into the period of the cybernetics controversy.[7]

The impact of the Zhdanovshchina was widely distributed throughout Soviet science, and some Soviet scientists paid for their "cosmopolitanism" and/or "agnosticism" with adversities in their careers and lives. However, the character of these adversities varied from science to science and from scientist to scientist. While some people were ignored after being attacked, others were urged to recant and contribute to the "new" Soviet sciences.[8] In a similar way the scientific impact of the campaign seems to have penetrated to varying degrees in different fields of science. One rough dichotomy can be made between the life sciences and the physical sciences.

In the physical sciences (mathematics, physics, and chemistry) the scientific consequences of the campaign appear to have been lighter and more transient. Perhaps the most important reason for this transient effect was the undeniable and constant demand for "materialistic" theories that would be as practically useful as those "idealistic" theories which were under attack. The substitute theories simply did not materialize.

The practical, utilitarian achievements of the natural sciences and their contributions to Soviet industrial and defense technology had been an important (and perhaps dominant) theme in the development of Soviet attitudes toward science. In the absence of useful, productive replacements for "bourgeois" theories (such as the theory of relativity and quantum theory in physics and resonance theory in chemistry), the Soviets often stressed both the continued use of these theories and the demand that they be somehow stripped to their "materialistic core." Therefore, attacks in these sciences were more often concerned with the interpretations of the consequences and bases from the viewpoint of dialectical materialism, rather than with altering or replacing their essential technical core.

The continued use of the attacked theories, even after assurances about their materialistic nature, produced further polemics during which some scientists were accused of merely "going through the motions" of embracing the philosophical-ideological tenets of the campaign.[9] In other cases, some scientists appear simply to have withdrawn from publishing for a time, in spite of repeated exhortations for them to admit their errors and continue contributing to their sciences. In any event, substitute theories based on "socialist science" were not produced, and in the long run the technical cores of these "hard" sciences did not appear to be significantly altered by the onslaught.

This situation can be contrasted to that in the life and social sciences, where alternative theories were advanced and gained a significant foothold in several fields of science. In this case, moreover, there were important mutual supports and interrelations among several fields. It was probably the systemic character of this ideological and scientific interpenetration which made the construct longer lasting and more profound than in the case of the physical sciences. It was also the systemic nature of the conglomerate which made it difficult to discuss new approaches in one field (for example, "biocybernetics") without sparking debates in the most diverse areas of the social and life sciences (for example, physiology, psychology, medicine, and so forth).

One of the cornerstones of this edifice of dialectical materialism and science was the theory of the origins of life advanced by the biochemist A. I. Oparin in 1936. The theory maintained that life originated from hydrocarbons and ammonia and water, from which the first proteins were formed. Proteins eventually combined to form so-called coazervates, which were the principal objects of investigation by Oparin and his students. These protein bundles exhibit the capacity for organization and can increase their stability by drawing upon selected portions of their environment. With the passage of time these bundles became progressively more complex until a "qualitative leap" was made to form the earliest living organisms.

At this point there already occurs a fundamental premise of Soviet biology, physiology, and medicine: the assertion that the very concept of life must consider both the organism and its surroundings. This statement was amplified in virology and cell development. In the former case there was the observation that the virus, as an intermediate form of life, lacks the means of metabolism when removed from its environment but can be regenerated when reintroduced into it. In the case of cell development there was the vilification of the German scientist R. Virchow (who held that cells can develop only from other cells) and the extraordinary career of Professor O. B. Lepeshinskaya. In the thirties she claimed the ability to produce cells experimentally from noncellular material, and her work was then criticized severely both in the USSR and abroad. However, after World War II she was supported by Academicians T. D. Lysenko and A. I. Oparin, and she was awarded the Stalin Prize in 1950. Lepeshinskaya was linked to Oparin by her work on the formation of cells from coazervates.

The theme of the union of organism and evironment was elaborated in considerable detail by T. D. Lysenko, who based his theory of the inheritance of acquired favorable characteristics on this equilibrium. Oparin, Lysenko, and Lepeshinskaya were all active in their careers in the thirties, but their contributions to "socialist science" were strongly underscored in the campaign of the late forties. In Lysenko's case, this occurred most strongly at the "historic August meetings" of the Lenin Academy of Agricultural Science in 1948, when he introduced the "reorientation" of Soviet biology.[10] Lysenko claimed to possess the capability of experimentally creating new species from old purely by the manipulation of the species' environment. His claim was tied to Lepeshinskaya's theory on the origin of cells from noncellular matter.[11]

The final ingredient in this congolmerate consisted of Pavlov's physiology, which had contacts with psychology and medicine that were later exploited in building this philosophical, ideological, and scientific system. As was the case with the above sciences, the dominant theme of Soviet physiology was to remove any gap between material and spirit, stressing the organism in its environment. The "orthodox apotheosis" of Pavlov's work in physiology occurred at the joint meeting of the USSR Academy of Sciences and Academy of Medical Sciences, which was held from June 28 to July 4, 1950. The principal speaker was K. M. Bykov, who spoke on "The Development of Pavlov's Doctrine."[12] Bykov's major charge was that Soviet physiology was currently under the influence of idealistic foreign theories that denied the dominant role of the central nervous system in the entire activity of the organism. He particularly attacked L. A.

Orbeli (who was succeeded by Bykov as director of the Academy's Institute of Physiology in 1950), P. K. Anokhin, and others.

The consequence of these "reorientations" in other fields quickly became apparent. In medicine, the "materialist concept of the unity of organism and environment" was evidently to become the central preoccupation of Soviet theory and practice in medicine. [13] In psychology and even pedagogy, the Pavlovian view of the central nervous sytem as a mediator between organism and environment was used to characterize speech and thought as a purely "material processes" (thus avoiding linking idealistic concerns with anything else). A link was even fabricated between Pavlov's research and Lysenko's doctrine of the inheritance of acquired favorable characteristics. [14]

In these ways the period of cultural chauvinsim witnessed the development of what appeared to be a distinct "socialist science" made up of elements from various life and social sciences. The body of theory was sufficient to make the fields mutually supportive, and the entire edifice was placed largely on the twin pillars of Sechenov-Pavlov in physiology and Michurin-Lysenko in biology. This edifice did not develop overnight but was based on work that had been done in the thirties or earlier (Pavlov won the Nobel Prize for his work in 1904 and died in 1936).

Whereas alternative descriptions of reality did not emerge in the physical sciences, in the life and social sciences this complex of ideas seemed to have a serious claim to be such an alternate. Furthermore, the complex was laced together with important strands from Marxism-Leninism-Stalinism and dialectical materialism. Finally, it was of enormous importance that this intellectual conglomerate not only had aspirations as a theory but claimed to be based on empirical investigations in the laboratory and also promised enormous (even miraculous) practical achievements. [15] Both of these latter claims were largely absent from the disputes in physical science.

Altogether this work formed a rather imposing knot, especially when the work and its sponsors also had the backing of the party leadership. It was evident that portions of the knot had to be untied and the "monopolization" of some of the disciplines had to be weakened if cybernetics was to find a place in the Soviet Union. Of even greater importance was the fact that, in the estimate of many Soviet scientists, these same tasks had to be performed if the "traditional" fields themselves were to be advanced (for example, if serious work was to be performed in molecular biology).

The "Stalinists" of Soviet science often chose to enhance their positions and arguments by posing scientific questions in the most spectacular political context of their time: the polarization of East and West that was occurring with the development of the cold war. As a consequence there may have been a tendency for others to analyze the developments of the Zhdanovshchina in their terms, seeing the system as an exaggerated

monolith to which the "Stalinists" clearly aspired. To do this is to miss the important point that, although the battles often emerged in global terms and ideological
conflicts, a great deal of the passion and importance arose from the fact that major
battles for control or independence were also being waged entirely within the Soviet
system. [16] Before examining some of the post-Stalin themes with regard for the
Soviet system of science, it is necessary to consider the character of the aspirant
monopolists themselves.

THE "MONOPOLISTS" OF SOVIET SCIENCE AND CYBERNETICS

The arguments about the "methods of cybernetics" were carried on most noisily in
those fields of Soviet science which were headed by officially recognized spokesmen
of "the only correct school" in that field. It was not true that "cybernetics" was
always responsible for the "new" arguments advanced, since in fact on some occasions
the attackers of the monopolies presented work that had been done before the con-
strictions of the Zhdanovshchina (merely offering them later as examples of "cybernetics and its applications" in a variety of fields). In addition, proponents of these
independent activities also argued their cases outside the context of the cybernetics
disputes.

What is often observed is the use of cybernetics as a part of a shield behind which
discussions could be held regarding the reestablishment of some individuals and their
"subschools" that had been eclipsed during the last years of Stalinism. In biology we
find "biocybernetics" and discussions of the merits of classical genetics, molecular
biology, and such figures as N. P. Dubinin; in economics there were discussions of
the normative role of economics in the Soviet socialist system, economics as an
"exact experimental" science, and the merits of such figures as Nemchinov, Novozhilov, and Kantorovich; in physiology and medicine there were discussions of the
work of L. A. Orbeli, electrophysiology, and the "reverse afferentiation" work of
P. K. Anokhin and others in the thirties; and so forth.

This is not to suggest that there were no new topics injected into these discussions.
It should be recognized, however, that the major source of passion in the disputes
stemmed from the interests of the defenders of "patents" in the "orthodox" system
of science, and from the desire to destroy some of those "patents" in an effort to
return to unmolested (and supported) work in "independent directions." If the term
"cybernetics" became increasingly vague in the course of these discussions, it was
partly just that it was distorted for the diverse purposes for which it was used.

The quest for this legitimation of subschools must be considered with simultaneous
demands for science to remain ideologically pure, and the continued tendency to

consider the planning of Soviet science along unilinear paths. These latter demands
in Soviet attitudes toward science have produced at times an environment that is
almost automatically hostile to subschools in some fields. For example, at the conclusion of the "Pavlov Sessions" in 1950, the joint decree issued specifically condemned the attempt to form schools and subschools in physiology which would thereby
be in opposition to "the general term of the Pavlov teaching."[17] There can be little
doubt as to whom Bykov thought fit to interpret precisely what the general trend of
the Pavlov teaching was, and it was that "patent" and the attendant centralization and
discipline that formed the major support of his "monopoly."

The decree from that conference partly reflected Bykov's denunciation of sundry
schools in Soviet physiology (including one that embraced the American "pseudoscience" of psychosomatic medicine), schools that presumably resisted the impact
of his current campaign. However, Bykov concluded his presentation with a statement
that revealed the tendency toward stagnation and dogmatism which was a consequence
of the establishment of single "official schools." He demanded that further research
to develop Pavlov's doctrine be performed strictly according to problems Pavlov
himself had raised, or on topics proceeding from the "essence" of his teaching.[18]

Finally, the campaign against independent schools and subschools was sometimes
similar to other charges against "familiness" in Soviet society, that is, against groups
of mutually supportive people who, by internal cooperation, could divert or resist
the direct impact of the central regime. For example, in his "Letters on Linguistics,"
Stalin attacked N. Ya Marr in ways that others used later in assaults on individuals
and their schools in physiology and biology. Marr was attacked not only for his theory
on the class basis of language but for having established an "Arakcheyev-type regime"
in linguistics which survived only by suppressing criticism within areas under Marr's
control.[19] Both the "Letters" and subsequent campaigns in other sciences depended
heavily upon the use of "criticism and self-criticism" as a device to break up these
mutually supportive groups and to guard against stagnation in Soviet science. In the
"Letters" Stalin also established a motto of enduring use in the years after his death:
that no science can develop without differences of opinion and the free clash of opinion
and criticism.

Obviously, these themes from the "coryphaeus of science" assumed different meanings at different times in the USSR, and many of them were later conveniently turned
on the very people who used them initially to establish their own "patents" in various
branches of science. In 1950, however, they were used to discredit and undermine all
but the "one correct path" in science. In physiology, they were used for Bykov and
against the "Beritov school," the "Anokhin school," the "Orbeli school," and others

whose alleged "feudal strivings" led them to engage in factional disputes over predominantly organizational issues (as opposed to scientific issues), thus inhibiting
the development of Soviet science.[20]

It would be a mistake to believe that this penetration of "socialist science" into the
life and social sciences was total. Evidence of continued resistance during Stalin's
life is necessarily indirect, but nonetheless is implied by repeated exhortations to
follow the new lines or in repetitions of the charges made at the original targets.
For example, an editorial entitled "Develop Criticism and a Clash of Opinions in
Science" appeared in the journal Medical Worker on February 6, 1953. The editorial
revealed that two and a half years after the Pavlov Sessions, the resolution to hold
discussions on higher nervous activity, its typology, and problems of genetics had
still not been carried out. Thus, it was not "surprising" that the most important
area of physiology had seen the fewest successes. Similarly, an exhortatory editorial
entitled "The Lagging of Psychology Must Be Overcome," in the Literary Gazette of
July 28, 1952, claimed that reticence in Soviet psychology had not been overcome
and psychologists remained uninterested in the demands of life in spite of appeals
from an "army" of Soviet pedagogues. Indeed, discrepancies between psychological
theory, Pavlov's doctrine, and Marxist-Leninist theory had only become "more apparent."

On the other hand, it is known that many Soviet scientists suffered damage to their
careers (and even loss of life) over issues related to the development of "socialist
science."[21] At least, it was obvious that there were several conflicting models of
science in the Soviet system, and it was this tense milieu into which cybernetics was
cast for its initial development in the USSR.

EARLY ARGUMENTS AGAINST CYBERNETICS

The impact of a "new science" upon any system is likely to create controversy, especially where those involved attempt to determine precisely what the proper relationships between the "new" and the "old" sciences should be. In the Soviet Union these
problems were particularly severe because science was planned along unitary lines,
and there was a tendency on the part of party leaders to confuse multiple approaches
to scientific problems with duplication of effort, loss of control, and imperfect coordination. In order to find a position in the USSR, cybernetics not only had to be seen
as a legitimate independent field to which funds could be devoted, but it also had to
find formal structural representation in institutes, councils, and other organizations
in which scientific progress is pursued.

Cybernetics was perceived in the USSR by its opponents as presenting both external and internal threats to the Soviet system of science. At the external level, cybernetics emerged from the West at a time when some Soviets were used to the idea of two fundamental and implacably opposed systems of science, the one socialist and the other bourgeois. Thus cybernetics was a priori hostile to "socialist science." The internal disputes brought forth two charges at the ideological-philosophical level: that cybernetics did not meet the canons of objectivity that were demanded by Soviet materialism; and that cybernetics was "mechanistic" and "reductionist" in its alleged attempts to "replace" some traditional sciences with the study of control systems. The first charge was used to question the legitimate existence of cybernetics itself, and the second was used to challenge its legitimate use by other disciplines.

The internal threats were largely concerned with the multidisciplinary nature of the new field of science, and what were taken to be its pretensions toward becoming a universal science with universal applications. Specifically, there was evidently the fear that with cybernetics several fields of science (for example, biology, physiology, sociology, economics) could be connected with one another by means of this "science of sciences" in a way that would be independent of the philosophical, ideological, and scientific activities carried on during the Zhdanovshchina. Since the security of the holders of monopolistic positions in Soviet science depended upon the continued recognition of their positions as "orthodox" interpreters of an already perfect ideology and philosophy of science, the specter of using new methods of research and independent means of integrating the sciences was a direct challenge to their authority and legitimacy. These people were no doubt the ones with the greatest interest in characterizing such challenges as being due simply to the hostile machinations of "bourgeois science." With time, it became increasingly apparent that they were also trying to maintain the inviolability of their exclusive preserves from the inroads of other Soviet scientists.

Academician A. I. Berg, one of the earliest and most enthusiastic supporters of cybernetics, clarified some of the political-scientific issues after cybernetics had won a place in the Soviet system.[22] Writing in 1961, he asserted that, while cybernetics had suffered from the errors of Western philosophers and publicists, that "unscientific shell" merely covered the useful and creative work going on among scientists, engineers, and economists working in cybernetics. He quoted A. Borin's article, "Science and Philosophy: (Kommunist, No. 5, 1960, pp. 100-101) to assert

"Attempts by any school of natural science to occupy an monopolistic position in science, to speak in the name of dialectical materialism and then to make use of this position

to 'unmask' scientists with other opinions as idealists or metaphysicists or reactionaries is nothing but nonsense and cannot be accepted. Monopoly will lead to stagnation in scientific life."

Berg immediately continues:

In this connection I wish to return again to the great problem of the application of
mathematics to certain sciences: biology, medicine, economics, law, and others.

Nonetheless, the arguments about cybernetics remained at the ideological and
philosophical level in the USSR for a long time. The Soviets probably paid a price
for keeping the dispute at the systemic level, since in doing so they made it extremely
difficult to handle the issues on a one-at-a-time basis within individual fields of
science. One of the reasons for this concern with the central questions of cybernetics
was simply that the confrontation became systemic in scope, in that some opponents
of the science wished to conserve the old system in every detail while others wanted
to alter or destroy major parts of it. Second, there is the familiar Soviet propensity
to regard all topics as massive, pervasive threats or massive, pervasive promises
at the most general level. Finally, carrying on the dispute at this level probably
allowed spokesmen access to the press without thereby revealing the details of disputes in individual sciences to people outside (and inside?) the Soviet Union.

Three general tendencies emerged in the attacks on cybernetics that were mounted
in the early arguments.[23] The first criticism was that cybernetics was "subjectivist"
and "idealistic." This criticism was connected with the assertion in dialectical materialism that there exists a material, objective reality that obeys objective laws,
neither of which depends in any way upon man's ability to perceive them. Complement-
ing this tenet is the notion that the universe, although infinite, is (in the limit) know-
able.[24] This criticism then lumped cybernetics with other theories of "bourgeois
science" which were then under (rather ineffective) philosophical attack in the USSR.[25]

One of the more coherent objections that finally emerged from this line of attack
was to the use of "functionalism" in cybernetics. Basically, this concerns the fact
that cybernetics (for example, in servomechanisms and control theory) examines only
the functions (transfer functions) of the system which determine its control characteristics. The major objection to such analysis in the Soviet context was that the functions of a system were not uniquely associated with a given material structure, that
is, the same control functions might be associated with systems of widely differing
structure (for example, the Soviet physiologist's brain and the Soviet cyberneticist's
computer).

The charge of "subjectivism" was associated with an inclination for the opponents
of cybernetics to "absolutize" the functions, little appreciating that the servomecha-

nisms game is played like other endeavors at modeling, that is, the analyst develops
the constituent elements of his functions from physical insight and trained intuition
until "adequate" agreement with empirical measurement is obtained. Ultimately the
discussions did involve the use of models and their gnosiological and epistemological
implications. These discussions first involved questions of identity versus analogy,
and even the legitimacy of using the power of analogy in scientific research. Second,
they involved cybernetics as perhaps the foremost science of "black boxes," that is,
substructures whose precise causal interconnections need not be understood so long
as the analyst can predict what the "black box" will produce as "output" when acted
upon by a specific "input."[26] This trend in the argument offered the opponents of
cybernetics the opportunity to link the new science with "Machism" and "positivism,"
two developments in the philosophy of modern science which had been attacked by
Lenin himself.

The second avenue of attack on cybernetics had to do with Engels' typology of
sciences that was based on the qualitatively distinct forms of "motion of matter" examined by the various fields. Engels' typology and associated dialectical laws of
quality and quantity were most often used to assert that cybernetics had no unique
subject matter that could be characterized by one form of motion of matter. Thus,
the argument ran, at best cybernetics could not be considered to be an independent
science with its own (unique) subject matter. At worst, it was a "reductionist pseudoscience" that attempted to analyze all possible systems ("even life and social systems") without regard for their qualitative differences.

We can examine Engels' typology and sample some of the subtlety of the arguments
by noting the following piece of intellectual judo which was obviously aimed at utilizing
the power of the typology against the opponents of cybernetics who proclaimed themselves the defenders of orthodoxy.

F. Engels proposed a classification including five forms of motion of matter (mechanical, physical, chemical, biological, and social). He pointed out that there are qualitative differences among these forms of motion. He also took note of the independence
of the lower forms from the existence of the higher forms, and emphasized that the
higher forms include the lower forms but do not replace them....Indeed, forms of
motion are distinguished neither by the masses and chemical compositions of the interacting bodies nor by their energies. Forms of motion are distinguished from one
another, in particular, by the level of the information processes. The natural science
characteristics of moving matter in each specific case are determined by definite
levels of organization, definite structures of interacting bodies, definite energy levels,
and definite levels of information.[27]

According to conventional interpretations of Engels' typology, it is generally not legitimate for a science to examine more than one qualitative type of motion, even though

the lower forms are contained within the more complex. Specifically, applying methods appropriate to a simple form (for example, mechanics) to a more complex form (for example, biology) is denounced as "reductionist." Battles over reductionism had previously occurred in the debates between the "Deborinites" and the "Mechanicists". In that case, the Deborinites advocated the inclusion of dialectical materialism as one of the first principles of scientific research. The Mechanicists, unenthusiastic about such an inclusion, held that all science could ultimately be reduced to physics and chemistry. Stalin, not yet the "coryphaeus of science," took no position. The Czech philosopher-scientist E. Kolman, in a role very similar to his later role in the cybernetics controversy, directed the factions to stop the sterile debates and get busy with the matter at hand: contributing to the development of the socialist state.

The most passion seems to have been generated over the problem of determining the position of cybernetics in the Soviet scheme for science, in particular with respect to the Engels' typology. Part of the problem was simply that it was impossible to alter cybernetics to make it fit the Procrustean bed of an outdated typology (indeed, the position of mathematics was always ambiguous in Engels' scheme). A second reason for the relatively long-enduring quarrels on this topic was that the Engels' typology was a weapon with the aid of which the opponents of cybernetics hoped to cut away its alleged universal aspirations, limiting it to at most automation and computer design. Since automation and computer design already had a history of some success and recognition in the USSR, had this movement to limit cybernetics been successful it is likely that cybernetics as a discipline would have been judged to be unnecessary and redundant.

These discussions about the use or abuse of the typology provided much of the vocabulary and structure about the application of cybernetics in three important areas: physiology, biology, and social science. In these fields the topics of greatest disagreement concerned the physiology of the higher nervous system, heredity, and the fields of sociology and economics. The charge was that cyberneticists were trying to use a methodology applicable only to simple systems in order to study systems of the motion of matter that were qualitatively more complex.

Although these discussions were carried out in the domain of Soviet philosophy (itself challenged to some extent by cybernetics), the arguments clearly gained momentum from the fact that they carried implications of great immediacy for the several scientific disciplines. Specifically, there was the threat of penetration of these fields by cybernetics or by physical scientists and mathematicians under the disguise of cybernetics. In the cases of physiology, biology, and economics, there was the addi-

tional problem that the disciplines to be penetrated were ones that, in the Soviet
Union, were then "political sciences" in the worst sense of the word.

As we have seen, the political content of these fields was underscored during the
period of cultural chauvinism which followed World War II. That period witnessed
the glorification of "official schools" in the life sciences: Sechenov-Pavlov in physiology (which overlapped into psychology, pedagogy, and medicine) and Michurin-
Lysenko in biology. Interest in the social sciences was restricted to the party, with
Stalin monopolizing any normative impact from their pursuit. Given the incredible
egotism of latter-day Stalinism, and the fervid faith that Truth was single-valued,
objective, and possessed uniquely by the "orthodox" leadership, it quickly became
heretical to suggest that "subschools" or "independent directions" had any right to
an existence.

At this point the philosophical objections to cybernetics begin to merge with the
third general line of attack, based on the fact that cybernetics was born "bourgeois."
As is the case with many a priori arguments, this line of attack did not generate
much of new interest in Soviet philosophy and was obviously mostly ideological in
nature. It did, however, raise some important side issues of importance to the
philosophy and politics of science.

The problem is that of constructing a stable system of science and scientific planning which can also allow flexibility and growth in important new areas of scientific
advance. There seems to have been a basic contradiction in the Stalinist system which
became apparent in many areas after his death. On the one hand there was at least
the partial development of a self-contained, ideologically interlaced system that was
based on "orthodox interpreters" whose task was to make certain that the only scientific questions raised were based on the essence of stale research. Contraposed to
this was the demand that Soviet scientists successfully compete in the world arena
and contribute usefully to the scientific and technological growth of the Soviet state.
Given both the rapidity of advance in world science and the systemic and conservative
nature of the Stalinist scheme, what Nathen Leites has termed the Bolshevik "ava-
lanche fantasy" may assume the character of the self-fulling prophecy.

With this thought in mind, and before considering the early arguments in the defense of cybernetics, we will now examine some of the post-Stalin changes in the
Soviet system of science. At the most important level, these were changes in the
leadership's perception of the proper role and performance of Soviet scientists.

CHANGES IN POST-STALIN ATTITUDES TOWARD SCIENCE

The years immediately following Stalin's death saw several important changes in
Soviet attitudes toward science. Three of these were of special importance to both
the content and tactics of the cybernetics dispute. The first general change concerned
a renewed exhortation to "let life judge" the effectiveness of science as well as the
correctness of scientific theory. On occasion, this appeal to utilitarian pragmatism
could be diverted and used to undermine the presumably orthodox ideological and
philosophical objections to cybernetics. The second theme, complementary to the
first, was centered on the exhortation to "learn from the West" in order to surpass
the West in peaceful coexistence and economic competition.[28] The third theme of
importance to cybernetics was the development of increasing reliance on automation
and related fields as basic means for increasing labor productivity and establishing
"the material-technical bases of communism."[29]

All three of these themes received some emphasis in the events of 1954—the year
that also saw the first public defense of cybernetics.[30] The strongest development
was the early campaign against dogmatism in science, often directed against Lysenko
and Bykov. This involved the new thoughts on "the unity of theory and practice" and
the demand that practical achievements (rather than adherence to dogma) become
the final test of scientific correctness. As the demand for practical contributions
from Soviet science to Soviet industry and agriculture became louder during the later
Khrushchev years, so the potential contributions from "independent paths" (whether
from the USSR or outside of it) began to attract greater interest. In the uneasy chase
after both achievement and reform, it seemed that no serious potential solution to a
practical problem would be turned down a priori on philosophical or ideological
grounds.

These themes were closely entwined with old arguments about subschools in Soviet
science which had been left unsettled—or rather which had been "solved" only by the
application of the famous "administrative methods" that characterized so much of
Stalin's era. The old monopolists were among the first to endure the pressure of the
drive to let "life and practice" be the criterion of scientific correctness. Plenary
sessions of the CPSU Central Committee in September 1953 and February-March
1954 condemned the losses incurred by the state as a result of the "dogmatic application" of agricultural theory. The ripples soon spread, and the president of the
USSR Academy of Medical Sciences censured the "intolerance of criticism" shown
by certain self-appointed monopolists in physiology and warned of the danger of
"vulgarizing" Pavlov's doctrines.[31]

The connections were quite evident in the important editorial entitled "Science and Life," which appeared in Kommunist in March 1954.[32] The editorial was a vigorous condemnation of dogmatisim as the "most dangerous sign of stagnation and backwardness" in science. Mobilizing the "free clash of opinion" in the struggle against "Arakcheyev-type regimes" in science, the article was especially critical of Lysenko's V. I. Lenin All-Union Academy of Agricultural Sciences.[33] The editorial condemned the practice of substituting administrative decrees for creative discussion of questions and the practice of "squeezing life into old formulas." The tests of "life and practice" were to be made the chief guardian against the danger of dogmatism in science.

In the course of considering the most pressing problems "advanced by life itself," the article lingered for some time on the social sciences. Philosophy, law, economics, and history were held up to especially severe criticism for being "divorced from life." Social scientists were urged to make practical contributions to current problems, abandoning time-worn dogmas and ancient history (which, of course, were the safest areas of activity during Stalin's rule). Of great interest was the point that these scientists were told (permitted?) to consider experience outside the purely Soviet context, and to study ongoing processes in the peoples' democracies as well as the capitalist countries and the colonies of the imperialists.

The general message of the editorial was one quite opposed to the voluntarism that marked Stalin's days. The suggestion was that the Soviet scientist, while considering pressing problems "raised by life" in the broad world context, could find "practical" solutions to "concrete" problems.[34] Earlier, these problems would have most likely been denied, or else "solved" by what Stalin's apologists refer to as "an exaggerated attention the adminstrative side of questions." Some of the problems mentioned were raising labor productivity, lowering production costs, increasing "complex mechanization," and studying the organization of work and the payment of labor (including the "consistent application" of material incentives).

The theme of "life and practice" was quickly followed by attention to scientific achievements in the West. S. L. Sobolev moved in this direction with his article, "A Scientist's Notes: On Scientific Criticism, Innovation, and Dogmatism."[35] Sobolev echoed the sentiments of the "Science and Life" editorial, broadening its targets by specifically including physiology as an area where dogmatism was a danger. Of greater interest, however, was his cautious focus on Western science.

Sobolev "solved" the contradiction between revisionism and dogmatism in science by pointing out that, while great Soviet predecessors were of great importance to Soviet advances, scientists of all countries were also highly valued (including Albert Einstein, who received attention from Lenin himself).[36] Sobolev asserted that Soviet successes in such high-priority areas as aviation, radiotechnology, and atomic physics had come forth only because Soviet scientists had been able to pursue new paths and open up new horizons in science, rather than merely commenting upon earlier work. He accused unnamed physicists at Moscow University of not contributing to progress in physics, even though they claimed succession from Lebedev, Stoletov, and Umov. Physicists who "rightly criticized" Einstein's "idealistic" world view should also have been able to see the rational in his concrete research.

Sobolev shifted also to denounce dogmatism in biology and physiology, where free discussion was controlled by the editorial boards of various journals. He condemned the habit of referring to any scientific opponents as "revanchists," "anti-Pavlovians," "Weismannists," and so forth. Advance in all areas of science could be best assured by checking ideas by "life and practice" rather than by adherence to "certain dogmas." Dogmatism could only harm the interests of the Soviet state, and the Party Central Committee taught Soviet scientists that dogmatism and stereotyped methods had to be exposed in order to turn scientists to the "real interests of Communist construction."

In the new struggles for Communist construction that marked the post-Stalin years, several old knots of influence began to be attacked by various groups of scientists. It is important to understand that there would have been at best a very tiny place for "cybernetics" in the USSR if the old official schools had been able to maintain their power and influence. It is also important to understand that had the "monopolists and dogmatists" been able to retain their position, there would have been very little room for such topics as mathematical economics, molecular biology, or electrophysiology, all of which have been linked at one time or another to "cybernetics" in the USSR. It was therefore no accident that the disputes over cybernetics were the hottest in fields of life and social science.

The quest for "liberalization" in the sciences produced strains similar to those encountered in other sectors of Soviet society during the Khrushchev years. The most obvious difficulty arose from the basic question that confronts all "reform from above": How does one contain the reform so that the legitimacy of the reformer is not undermined? In the USSR, there was the additional problem of preserving party and government control over the process of reform itself. Beyond this, there was the problem that the act of reform had to take place in a system in which there were still "Little Stalins" whose influence was in jeopardy. In such a system, the problem of producing meaningful change without recourse to coercion and "adminstrative methods" was a difficult one. As a result of these contradictions, the demand to learn from the West

and extract maximum utility from independent paths in research was accompanied
by setting limits beyond which basic aspects of Soviet society could not be questioned.[37]

There were two major results from these campaigns in the mid-1950s. First,
there was a clear tendency to extend a sometimes cautious legitimacy to independent
schools of thought, providing their arguments were sufficiently contained and practical to be regarded as nonrevisionist and potentially profitable to the state. Second,
there was an increasing call from various party spokesmen in science for cooperation between scientists and philosophers in order to assure the purity of new investigations. The latter activity was evidently resisted by philosophers and scientists
alike, and the ideological aspects of social and technical innovation seemed more
difficult than the leadership initially supposed.

By 1955 the explicit advocacy of "independent schools of science" had emerged from
the earlier comments on "life and practice," the demand for free clash of opinions,
and the campaign against monopolists of science. The advocates tended to be cautious,
no doubt partly because the old monopolists were by no means out of the picture. In
spite of the difficulty in which Mstislavsky and others found themselves, Dyachenko
went on later in the year to define pedantry and dogmatism as the "main enemy" to
discussion and advance in economics.[38] He gave as the reason for the seriousness of
the problem the new demands for economic competition with capitalism which had been
placed on the peaceful coexistence movement.

If the new international line stressed "peaceful coexistence," a similar development
seemed to pertain to intra- and interdisciplinary disputes in Soviet science. There was
little evidence of a post-Stalin purge in Soviet science, and little evidence that the "new"
subschools were aspiring to monopoly positions over the bones of their monopolist predecessors.[39] Rather, this was a time when representatives of all Soviet fields (physics,
chemistry, mathematics, biology, physiology, and others) were called upon to work out
their differences in a "comradely way," presumably paying close heed to the practical
needs of the state and the progress of Soviet science. Even Topchiyev extended a somewhat grudging recognition of the propriety of Soviet "schools" in order to provide supportive conditions to great scientists and their pupils. However, he also stated that
such schools could avoid being mere fads only so long as they were based on "Marxist-
Leninist methodology" and observed "party principles" in science.[40]

The theme of "peaceful coexistence" between subschools enjoying some autonomy
was ancillary to the second major development of the time: the demand that scientists
and philosophers work more closely together to assure both the extraction of maximum
usefulness from the Soviet system. This call also made it clear that although "revi-

sionism" would not be tolerated, at the same time mere adherence to timeworn dogmas could not be the major criterion for scientific excellence. The result of the campaign was to make an ideological attack on a field of science both more difficult to mount and harder to sustain. [41]

To some extent, Soviet philosophers were made the scapegoats for various lags in science and technology. In 1955 A. A. Maximov, who had earlier accused Fok and others of trying to drive a wedge between scientists and philosophers, was dismissed from the editorial staff of Questions of Philosophy under the same charge. In time, unnamed philosophers were increasingly accused of "nihilistically rejecting" advances in the natural sciences which they did not understand, and unnamed natural scientists were criticized both for their alleged "positivist" leanings and their inadequate interest in understanding Soviet materialistic philosophy. [42]

The international and national problems in the reform were seen in the Soviets' regard for Western work. In the April 15, 1956, issue of Pravda, Einstein was described as a "natural materialist" in his work, although in philosophy he vacillated between "a timidly expressed materialism" and idealism. [43] Topchiyev, in his busy report on Academy activities in 1954, asked that Soviet scientists "take everything positive" achieved in world science and technology, while exposing its reactionary, bourgeois ideological orientation. According to an editorial in Kommunist, the party carried out "the only correct line" of alliance with scientists from capitalist countries. These scientists, although often under the influence of idealism, were "basically materialistic in their research. [44] Even under capitalism, where science was held in chains, there were "progressive" tendencies struggling against reaction, and it would be absurd to ignore the advances in world science. At the same time, it was necessary to prosecute vigorously the struggle since the "positivistic points of view" of foreign scientists were not without a "certain influence" on Soviet natural scientists.

It is likely that some scientists and philosophers resisted the demand that they resolve the questions and end the polemics which had been so much a part of the Zhdanovshchina period. Some articles were principally aimed at certain dogmatist philosophers, and advanced the promises of pragmatic utilitarianism and practical benefits from scientific advance. Other articles continued to emphasize the dangers of the "positivist preachment" and alleged attacks on materialism under the guise of "experiment," "fact," and "empiricism."[45] At least as early as February 1957 an article appeared calling for fuller cooperation and coordination between social scientists, natural scientists, and philosophers, mentioning a national conference on philosophical problems of natural science to be held the coming June. [46] In September 1957

Topchiyev called for cooperation between physicists, chemists, biologists, and philosophers, and stated that convening a national conference on philosophical problems of modern science would be helpful.[47]

The apparent lack of resolution of these relatively durable tensions in Soviet science and philosophy was the setting for the 1958 Conference on Philosophical Problems of Natural Science. After a reluctant beginning, the conference was held in Moscow on October 21 to 25, under the joint sponsorship of the USSR Academy of Sciences and Ministry of Education. The presentation of cybernetics at this conference is examined in the next section. Here, it should be noted that in the October 31 issues of Pravda and Izvestia the Presidium of the USSR Supreme Soviet announced that S. L. Sobolev (who presented cybernetics at the conference, along with A. A. Lyapunov) had been awarded the Order of Lenin on his fiftieth birthday for his work in mechanics and mathematics.[48]

THE DEFENSE OF CYBERNETICS

Before turning to the explicit defenses of cybernetics that emerged from 1954 to 1958, it is useful first to examine the third relevant theme in Soviet attitudes toward science —the growing involvement with, and faith in, the spectacular "materialistic expressions" of the new field of computers and automation. Evidently, developments in the field of automation and remote control ("telemechanics") continued to develop through the years of the controversy, although the pace had become more rapid in the years after 1958. This progress (later partly included under the rubric of "technical" or "engineering" cybernetics) provided a considerable resource for the defenders of the new field.

Soviet activity in automation and remote control began quite early. The First All-Union Conference on the Theory of Automatic Control was held at the time of the establishment of the USSR Academy of Sciences' Institute of Automation and Telemechanics in 1939. The second such conference was held in 1953. The development of digital computers preceded the public debate about cybernetics in the USSR, and the Academy's Institute for Precise Mechanics and Computation Technique was established in 1950. The first Soviet digital computer, the MESM, was developed at that institute by S. A. Lebedev, who in 1952 developed the larger and faster BESM-1 machine.[49] Given the size and complexity of BESM-1, it is likely that the decision to build it was probably taken by the USSR Academy in 1950 or 1951. In 1956 A. N. Nesmeyanov (then president of the Academy) spoke of the Academy's initiative in producing BESM.[50]

Only months after "Materialist's" vitriolic attack appeared, it was obvious that
the Academy was on the path of considerably widening its computer capabilities.
This interest was vividly expressed in the important and long overdue Academy
elections in October 1953. In the list of vacancies reported in Izvestia on July 22,
it was announced that two computer specialists would be elected corresponding
members in the Physics and Mathematics Division, while four corresponding members and one academician would be elected to the Technical Sciences Division in the
fields of radio engineering, electronics, automation, and telemechanics. Both the
Academy president and secretary gave special emphasis to this new direction in
Soviet science, and Topchiyev linked it to the attention given at the Nineteenth Party
Congress to automation and telemechanics "characteristic of an era of communist
construction."[51] On September 3, Izvestia added 3 academicians and 4 corresponding
members to the list of vacancies. The election selected 51 academicians and 148
corresponding members, among whom there were 5 academicians (including S. A.
Lebedev) and 9 corresponding members in the above fields.

In December 1955 A. N. Nesmeyanov published an interesting article in which he
forecast that the most promising technological breakthroughs would occur in atomic
energy, semiconductors, and computers. He stressed that the most promising areas
of science were those "at the juncture" where the methods of one science could be
used in another.[52] As an example, he discussed the use of modern electronic equipment to study the brain's activity and the conditioned reflex "as established by I. P.
Pavlov." He also stated that Soviet computers would "find their first application" in
statistics and planning, and later in information retrieval and the control of individual
production processes. Most important, such machines could increase the productivity
of the mental processes themselves.

In his speech at the annual general meeting of the Academy in 1955, Nesmeyanov
revealed that the Academy Presidium had reserved for its direct supervision eleven
of the most important problems of Soviet science. The solution of these problems
would "open new horizons" and provide "qualitatively new possibilities." Four of these
problems (and their associated academy institutes) were high-speed computing machines and equipment (Institute of Precise Mechanics and Computation Technique),
semiconductors and their technical applications (Institute of Semiconductors), enhancing the effectiveness and reliability of radiotechnical installations and their equipment (Radioelectronics and Electronics Institute), and securing the development of
automation and remote control in production processes (Institute of Automation and
Telemechanics).

During the next three years (1956-1958) the stress on the development of automation and computers grew continuously, receiving vigorous interest in the wake
of the Soviet space program. Perhaps the most important pronouncement in these
areas was made at the Twentieth Party Congress in 1956. Later the proceedings of
this conference were quoted extensively to justify both the products and pursuits of
cybernetics. The directives of the congress asserted the fundamental importance
of increased automation of production as an important way to increase labor productivity and industrial abundance. It was forecast that automation would change the
character of the worker's labor, making it more like the labor of technicians and
engineers. In addition, it was asserted that under socialism the increased productivity would be used to raise the workers' cultural and technical level. Finally, the
general problem of studying the maximum effectiveness and optimum utilization of
new, automatic equipment was posed by the Party for the economists. [53] The drive
toward automation embraced an exceedingly wide range of science and industry and
was quite consistent with the demand that science and technology become more intimately connected with the productive bases of the Soviet system.

One of the earliest public defenders of cybernetics in the Soviet Union was the
philosopher-mathematician E. Kolman, who began the defense of the science at least
as early as November 1954. [54] At the time his "What Is Cybernetics?" was published,
the demands were current that Soviet scientists use the results of world science to
solve the problems that had been "raised by life itself" for the builders of communism. Kolman's first paper emphasized the utilitarian promises of the field, while
positing the realistic possibility of stripping it of any of the trappings of bourgeois
ideology. Furthermore, he charged that "certain philosophers and scientists" in the
USSR had rejected the field without knowing anything about it aside from what had
been said by some "bourgeois scientists and publicists." [55] Therefore, the implicit
argument ran, the task was simply to explain patiently and correctly the principal
features of the science so that its opponents would be suitably enlightened and abandon
their past errors. It was simply "nonsense" to believe that the enemies of the Soviet
Union had devoted so much time and energy to the field simply to discredit Pavlov and
"draw idealism and metaphysics into psychology and sociology."

Kolman tried to sidestep controversies going on in formal logic (as opposed to
dialectical logic), merely noting the practical successses which had accrued to formal
logic in several branches of mathematics and computer design. He denied that any
field was trying to "take over" another field, arguing instead for the power of analogy
in many fields of science and for the promise of success in using methods developed

in one field for research in another. For the social scientists in his audience, Kolman observed that Marx himself wanted to apply mathematical methods to the analysis of social and economic phenomena. "Why," he asked, "should quantitative methods yield practical results in every other science and not in psychology?"

Kolman had occasion to present a longer defense after the Twentieth Congress in 1956.[56] He changed some of the major points of the argument to show a greater tie with the campaign to "let life judge" scientific correctness, and relied for support on the detailed specification of the tasks of automation which emerged from the directives of the Congress. Interestingly, his examples of the achievements of cybernetics in industrial and defense technology were largely taken from the United States. While the Soviets lagged behind in the civilian applications of automatic machinery, "the American press writes that Soviet armament is not falling behind the American in this respect." Bringing in themes familiar from other disputes, Kolman asserted:

The nihilist attitude displayed by some of our philosophers toward cybernetics is as harmful as the nihilist attitude toward the theory of relativity, the application of quantum mechanics in chemistry, the study of heredity using physics and chemistry, or mathematical logic. Nihilism, the rejection of scientific achievements on the grounds that the ideology enveloping them is antiscientific, holds down the growth of our science and technology. . . . And confusing the philosophic view of the naturalists of capitalistic countries with their political aims, counting them in the camp of reaction only on the grounds that they are idealists, can only alienate from us many sincere friends.

Kolman concluded his presentation with a statement of faith which was endlessly repeated in the coming years. This was the assertion that only in the rational, planned Soviet system could cybernetics rise to its true potential. Somewhat circularly, it was reasoned that cybernetic technology would make possible the grand features of the Soviet plan, that is, the elimination of the difference between mental and physical labor, the shortening of the working day, and the creation of abundance and cultural growth of the nation.

Several of the themes and tactics used by Kolman were further developed by S. L. Sobolev and A. A. Lyapunov in 1958 at the First All-Union Conference on the Philosophical Problems of Natural Science.[57] We have already noted the pressures building toward convening the conference and the hope that the conference would produce a productive cooperation (or at least a halt to sterile battles) then being waged between factions in science and philosophy. Given these expectations, it was mildly surprising that Sobolev and Lyapunov largely chose to ignore philosophical problems in their exposition. On occasion, these were almost superciliously dismissed:

We admit that we do not even understand some of these (philosophical) questions in relation to cybernetics. Sometimes we are forced to hear from the camp of the philosophers the demand that one must "materialistically explain the philosophical meaning of electronic computers." We admit that we do not quite understand how one can explain philosophically the electronic computer....

It seemed to us that if we would illuminate in more or less complete form...the question of what modern cybernetics actually is...that many of these misconceptions (rather sad ones, about which we have read in the pages of our press, and which we have often heard) would disperse automatically.

In short, their stance was the same as Kolman's: treating the controversy as the result of misunderstanding and misinformation, mostly on the part of philosophers. The authors shunned considerations of philosophical questions (including the definition of the field) but nonetheless insisted that concern with such questions should be postponed until more data were available, and in any case were of more interest to technical specialists than philosophers.

The authors discussed three "aspects" of the field: control theory, information theory, and the theory of algorithms. In presenting examples, they chose to select almost entirely applications in military and space technology. They also identified the field much more closely with general computer use, a device of later polemical use for the field's supporters.[58] Having anchored cybernetics to computer development and application, as well as to high-priority areas of defense technology, the authors launched into an attack on biologists of Lysenko's persuasion.[59] They argued that the inheritance of acquired favorable characteristics requires a questionable path of information flow from descendant to parent.

Had this particular condition been taken into consideration, not a single scientist would have proposed the idea of adaptive heredity or a directed evolution independent of selection.

Not only were some biologists getting wrong answers, they were not even asking the right questions.

It remains a fact that the flow of hereditary information, traveling from an organism as a unit to its embryonic cells, lies unknown and none of the followers of Lamarck asks the question of how to uncover it. On the other hand, the data of classical genetics correspond fully to ideas proposed by cybernetics.

The propriety of applying the new science in these previously sacrosanct fields occupied a good deal of the conference discussion, even being included as "microcybernetics" in the paper by V. M. Engelgart and G. M. Frank, "On the Role of Physics and Chemistry in the Investigation of Biological Problems."[60] Furthermore, the objections to cybernetics that were voiced seemed to flow into one another in an entirely illogical way unless we understand that what was being threatened by cybernetics (and other sciences) was the complex of life and social sciences and ideology which had come together in the period of the Zhdanovshchina. We will consider the expansion of cybernetics in the next section. Now, however, it is useful to summarize some of the major objections raised to cybernetics and the ways in which they were met during this first period. One can group arguments for and against cybernetics with relevant changes in party attitudes toward science which were occurring at the same time. There are three such groupings.

1. First, one of the major arguments against cybernetics came from the domain of philosophy of science. This was the accusation that cybernetics did not meet the canons of objectivity that had been established in Soviet dialectical materialism. Cybernetics was condemned as "mystical," "idealist," and "subjectivist" both because it presented categories beyond those accepted in Soviet materialism (that is, information) and because as a science of "black boxes" it suggested one did not have to know the detailed causal connections within subsystems in order to perform useful scientific analysis. In this way, cybernetics was linked to the old philosophical enemies of Machism, positivism, and pragmatism.

Within the constraints of Soviet dialectical materialism, these arguments against cybernetics were quite reasonable. The supporters of the science usually refused to quarrel with the philosophers on these issues (beyond saying "it isn't so" and calling their opponents "dogmatists" who stood in the way of scientific progress). Rather, they supported their arguments with clear appeals to utilitarian pragmatism. This change in "ground rules" for the argument made the philosophical objections seem impertinent and scholastic. Furthermore, the change was consistent with the party demands that science contribute concretely to solving Soviet problems.

This defense on the basis of utilitarian successes was to a great extent aided (and perhaps even made possible) by the new emphasis on "life and practice." The dominant message from the campaign was that simple adherence to "orthodoxy" would no longer be the main criterion for a "correct direction" in any science. Soviet scientists who promised practical achievements would be given the right to pursue their independent paths, subject to the limits of "revisionism."

2. The second connection in the cybernetics disputes concerned the accusation that cybernetics was "reductionist," that is, trying to study complex forms of the motion of matter with methods suitable only to simpler forms. This objection had two major components. One was philosophical and dealt with Engels' typology, which classified sciences according to the types of motion they studied. (The general spirit of the typology was occasionally conveyed by quoting Marx's comment that the key to the anatomy of a monkey is the anatomy of a man.) The second component was political-scientific, in that the accusation of "reductionism" was used by the "official schools"

and their spokesmen to keep mathematics, physics, and chemistry out of fields such
as physiology and biology.

On the philosophical level this attack was often concerned with the "man-machine"
problem (that is, can machines think?). Here E. Kolman erected the perfect defense
to meet the perfect offense. His argument was that if the opponents of cybernetics
could base their "reductionism" charges on the classics and Engels' typology, the
supporters of cybernetics could study "machine thought" on the equally orthodox
position that man and machine are a part of the same material universe and must
obey the same laws of natural science. Otherwise, his opponents would have to attribute "magical" properties to life, committing the idealistic sin of "neo-vitalism."

At the political-scientific level, the supporters of the field tended to reduce the
proclaimed domain of cybernetics, and to argue for their science from the strong
technical core of the discipline and its concrete contributions to automation and
computer development. However, the same spokesmen who reduced the domain
simultaneously demanded that the precise definition of its scope be left free from a
priori pronouncements from outside the technical discipline. They thus avoided the
danger of being "defined away" as mere automation. At the same time they argued
that control theory and compter design were merely trying to perform research on
the basis of analogies between control systems in life and technology, not establishing
identities between them.

While this aspect of the argument was being pursued, the party was expressing
greater interest in computers and automation for the assistance they could offer the
builders of communism. This assistance ranged from increasing labor productivity
during the period of peaceful coexistence and economic competition to actually trying
to use computers for the planning and direction of the national economy.

3. The third major objection to cybernetics concerned merely its birth in the West
at a time when Soviet men of influence were accustomed to the notion of "socialist
science" wrestling to the death with "bourgeois science." For these critics, the condemnation of the field as a "bourgeois pseudo-science" should have been enough to
banish it from Soviet soil.

The response of the supporters of cybernetics to this allegation was to work for
the "secularization" of the science or, in less frequent cases, for the establishment
of a Marxist-Leninist foundation for it. In this connection they tended to dissociate
cybernetics for the West and from Wiener. This was achieved both by adopting a
Marxist reading of the history of science—that science serves the cause of production
and economic development—as well as by pointing to Soviet or Russian contributions
to the field. In addition, there was the beginning of the idea that a man's scientific

work could be separated from his philosophy, ideology, and politics. This suggestion was alien to the spirit of "party mindedness" in all things but quite at home in the secularization of other fields of science which was proceeding in the mid-fifties (for example, in relativity theory and quantum mechanics).

While this contraposition was being made, there occurred the party campaign to use the products of new developments in Soviet and world science for the construction of communism and competition between the two world systems. This campaign was undertaken in a thoroughly Leninist, dialectical way, and exhortations to use the scientific achievements of the West existed side-by-side with demands that scientists and philosophers work together to construct Marxist-Leninist foundations for the advances, ensuring the inviolability of the system by continuing the ideological struggle. Nonetheless, as two Soviet scientists pointed out, within a decade no one continued to speak of "bourgeois physics," "bourgeois chemistry," or "bourgeois physiology," and to continue to speak of "bourgeois biology" was at best a deplorable holdover from the "cult of the individual."

A turning point in the disputes clearly had occurred by 1958, at the time of the First All-Union Conference on Philosophical Problems of Natural Science. Wiener's Cybernetics was translated into Russian in that year, and the occasional journal, Problems of Cybernetics began to appear under the editorship of A. A. Lyapunov. Within a short time the Scientific Council on Cybernetics of the Presidium of the USSR Academy of Sciences was established with A. I. Berg as its chairman. In the following years cybernetics expanded into some of the very areas that had been denied to it by its most vigorous opponents. By 1964 the Scientific Council on Cybernetics had sixteen sections, evenly divided between physical-technical and biological-social aspects of the science. The latter eight concerned economics, biology, psychology, linguistics, bionics, philosophy, law, and the theory of organization.

The sudden proliferation of articles on cybernetics and biology, physiology, economics, and so forth, suggests that what may have occurred was the quick publication of works that had been accumulating for a long time. In part, this was because the period of expansion assisted in the quest for the reestablishment and recognition of old and new "subschools" in several sensitive areas of science. This is not to say that new technical contributions from cybernetics were entirely absent in these new areas but merely that there was more going on than just the study of cybernetics and the pursuit of fashion. We will argue that there was a mutuality of interest between cybernetics and subschools in other fields which accounted for the quick establishment of the discipline in areas that had been so recently closed to it. We will examine some of these mutual interests in the next section.

THE EXPANSION OF CYBERNETICS

It is quite likely that the coalescence of interests which occurred between spokesmen
for the party, the life sciences, and the social sciences accounted for much of the
enthusiasm that surrounded cybernetics in the early sixties. We are not asserting
that there were no scientific contributions accompanying the developments of this
time but only that these interests in "the politics of science" (and the science of
politics) contributed much to the aura of fashionability which surrounded cybernetics
at the time.

It is quite likely that the expansion of cybernetics came after the establishment of
these "interest groups" in Soviet science. It could be used by various groups as a
way of supporting various sides of old and new arguments that were going on at the
time, getting them into the public domain.

With regard to the Party's interest, we have already noted the development of the
themes of "life and practice," using the products of world science to build communism
in the USSR, and utilizing automation to increase labor productivity. These themes
continued into the sixties, and it is quite possible that Party approval of cybernetics
was granted to the field as a science of automation alone. In addition, cybernetics
played a large role in Khrushchev's plans for peaceful coexistence and economic
competition with the West, both through its contributions to industrial technology and
in its value as a "showpiece" of Soviet scientific achievement.

The recognition which had been extended to the importance of automation at the
Twentieth Congress in 1956 had been taken by the supporters of cybernetics as an
indication of approval for the field. At the Twenty-First Congress in 1959, Khrushchev spoke both of lagging Soviet labor productivity and the need to overtake the
United States. [61] He also spoke on the Seven-Year Plan to achieve this, adding that
the goal could be achieved only with complex mechanization and automation. Automation had already been linked to such major communist goals as the elimination of
the differences between urban and rural life, and between physical and mental labor.
At the July 1959 Plenary Session on Automation Kosygin (then vice-chairman of the
Council of Ministers and Gosplan chairman) tied automation and speedier scientific
progress to the establishment of the material-technical bases of communism and to
the Soviet victory in economic competition with capitalism. [62]

By 1961 Khrushchev began to speak for cybernetics explicitly. At the Twenty-Second
Congress, he again emphasized the need to utilize more quickly and thoroughly all
the discoveries of Soviet and foreign science and technology, as well as the need to
accelerate complex mechanization and automation. He also asserted that Soviet scientists were "widely celebrated" for their achievements in physics, mathematics, and

cybernetics. The 1961 Party Program was altered from its draft in such a way as to make the portion dealing with cybernetics both more assertive and broad.

In the 20-year period integrated automation of production will be carried out on a large scale, with an ever growing shift to fully automated shops and enterprises ensuring high technical and economic efficiency. The introduction of highly perfected automatic control systems will be accelerated. Cybernetics and electronic computers and control systems will be widely applied in production processes in manufacturing, the construction industry and transport, in scientific research, in planning and designing, and in accounting and management. 63

If Khrushchev was interested in promoting the development of cybernetics, it should come as no surprise that there were spokesmen in the technical field who were interested in promoting Khrushchev's programs. Berg, in the editor's preface to the first volume of his Cybernetics at the Service of Communism, pointed out that work on the book had been under way since 1959, and its publication was timed to coincide with the convocation of the Twenty-Second Congress.

With regard to the threats and promises of automation, Khrushchev and various spokesmen for cybernetics quickly adopted the simplistic line that it was bad in the West and good in the USSR. In the West, automation would serve "bourgeois goals" and lead to further exploitation and unemployment; in the Soviet Union, it would be subservient to the "communist goals" of shortening the workday, decreasing the difference between intellectual and physical labor, and making urban and rural life more alike. On August 21, 1959, Izvestia assured its readers that, even with the great increases which would come from the Seven-year Plan, in 1965 jobs would be found for "all who wished to work."

Several of the spokesmen for cybernetics also made supportive comments regarding Khrushchev's "work ethic." For example, in discussing his 1958 educational reforms, Khrushchev denied that the imposition of work experience before higher education and again before graduate school was due to any labor shortage. Rather, it was to strengthen the ties between the school and life, for the moral and educational improvement of the people. In 1961, Kolman felicitously observed that since the work week would be shortened and physical labor minimized, all people should learn and participate in "productive work" in order to become conscious, disciplined members of society. This was the reason for uniting education and physical labor in the educational reforms, while at the same time creating the "material and technical bases of communism" with general automation. [64]

Party interests in the new science went beyond "mere automation," and there were many pronouncements on the potentials of cybernetics in optimizing the control and management of the planned socialist state. In 1961, Berg hailed the administrative decentralization which followed the February 1957 Plenum as "completely justified" and determined by the scale of production and technology which then pertained. He also predicted that advances in computers and automation would bring radical reforms and changes in the automation of new systems of planning, accounting, and statistics.[65]

Similar interests were expressed regarding what came to be known as "scientific management," a development of enduring interest whose spokemen have included both Kosygin and his son-in-law, D. Gvishiani. In the article on cybernetics in the 1962 Philosophical Encyclopedia, the importance of social control processes and the "scientific organization of labor management" were stressed. These involved systematic studies of the best way to reach social goals with minimum expenditures of time, effort, material, and energy.[66]

At the same time these interests between party spokesmen and cyberneticians were developing, arguments about subschool autonomy in Soviet science continued. In the published discussions of the Draft Party Program, D. Blokhintsev warned of the dangers of planning science along unitary lines, and the risks of giving a single school a "patent" on the truth of its own position.[67] He argued that scientific research was so complex that one could not afford to ignore any avenues of approach to problem solving, and emphasized the importance of broad, free discussions of scientific questions (which was included in the Party Program). Blokhintsev concluded by pointing to the imminent scientific revolution in biology, "at the juncture of biochemistry, cybernetics, and genetics."

Open discussions began to be reported concerning the impact of cybernetics on the life sciences in the USSR, particularly in biology and physiology. These discussions involved a wide spectrum of disciplines, including psychology, pedagogy, and medicine. On occasion, those speaking temporarily for cybernetics were at the juncture of these many activities.[68] In the course of these discussions, the participants used the familiar tactic of attacking monopolization of the field while associating various old and new lines of investigation with both cybernetics and the Party's expressed interests in the new field.

In 1961, an All-Union Conference on the Application of Mathematical Methods in Biology was held at Leningrad State University. An Izvestia article by V. V. Parin linked biology and the Soviet space program, and soon articles on bionics (the study of control systems in living nature for application in technology) began to appear. The latter were often accompanied with a stress on the military payoffs in bionics, and the practical American successes in this field which appeared to be "divorced from life."[69] In Pravda, April 16, 1962, it was announced that at the Central Committee's Plenary Session on Agriculture, T. D. Lysenko had asked to be relieved of his duties as head of the V. I. Lenin Academy of Agricultural Sciences (for reasons of health). This announcement coincided with the general meeting of the Biology Division of the USSR Academy of Sciences, which was devoted to the biological problems of cybernetics.[70]

In the papers that appeared after the conference, genetics and molecular biology had a very prominent place among the topics of interest to both biologists and cyberneticists. For example, N. M. Sisakyan observed that ribonucleic acids performing the information transfer of heredity had become of great importance to biology, medicine, and cybernetics (Watson and Crick received the Nobel Prize for their work on the subject in 1962). He warned that the main enemy of scientific progress was conservatism, a "slavish regard for outmoded traditions that shackles the creative potential of science."[71] In the same collection, Berg wrote on "The Science of of the Greatest Potentials." He pointed to the recent achievements of biologists, physicists, and mathematicians in biochemistry, some of which had been reported in 1961 at the International Conference on Biochemistry in Moscow. Berg stated that these investigations were called for in the Party Program which (no doubt to the embarrassment of some) stated: "Pursue more widely and thoroughly the Michurin trend in biology." Berg argued that it was merely a question of allied sciences helping "the heirs of Sechenov, Mechnikov, Pavlov, Bekhterev, Pryanishnikov, and Michurin" to find the "most effective means" of taking the lead in this branch of world science, "as demanded by the Party Program."

The fact was that there were still great problems in both the science and politics of biology in the USSR. Parin dealt with some of these in his paper in the same collection from _Nature_. He asserted that the period of great "emotional excitement" over cybernetics had passed, since "most" biologists and physicians recognized the field as a necessary element in both research and education. However, an enormous amount of scientific work lay ahead, and in that connection:

The most important element in this effort is the creation of the conditions for continuous creative cooperation among the representatives of the various fields....Nor can there be any doubt that a considerable portion of this work is many times greater than the visible portion, just as is the submerged portion of an iceberg. For some time it will remain hidden from the eyes of outside observers, and will become visible only in years to come, when mankind will be able to make actual use of major theoretical and practical achievements of medical and biological cybernetics.[72]

It is hard to know precisely what Parin's "iceberg" refers to, although it may been merely a warning not to attribute a lack of immediate success to the laziness or incorrectness of Soviet specialists working in the new field. On the other hand, he may have been referring to continued difficulties in working and especially in publishing in a still controversial area. In 1964 Parin called for the development of bionics and cybernetic studies of the brain, pointing out that there had been at least ten bionics conferences in the United States alone. [73] Nevertheless, a Western observer noted:

The Party lifted the ban on publication of works written by geneticists opposed to Lysenko, and the results of over a decade of silent research came to light. [74]

The problems of overcoming the effects of Lysenko's "monopoly" have evidently been quite difficult. A perhaps unintentional description of the difficulties was provided by Academician N. Semenov and A. Malinovsky in their article "The Relations Between the Exact Sciences and Biology." [75] The example with which they chose to demonstrate the usefulness of cybernetics and biology was the phenomenon of "life waves" as negative feedback. These population waves arise when the relationships between predators and their prey depart from equilibrium, they sometimes "damp out" very slowly. By the time Lysenko departed the scene, only a part of the problem was to attract talented students to an important field that had been dangerous during his rule. The equally difficult part was to produce an entire generation of teachers of biology, who would otherwise have emerged from the "lost generation" of biologists of the forties.

The development of cybernetics and physiology pursued a course rather similar to that of "biocybernetics." Open discussions came to involve work that had been done before the "Pavlov Sessions," new activities in the study of physiology, accompanied by a wave of articles dealing with "new" problems. [76] There was also the tendency for the discussion to spill over into the related fields in the life sciences.

Part of the excitement in physiology was simply that this was the major area in which the "man-machine" problem was fought out, that is, the charge that some physiologists and cyberneticists were reductionists who were trying to replace (or identify) man and his thought with machines and computers. These arguments were quite complex, involving not only the old "dogmatists" but also sharp divisions of opinion between physiologists, psychologists, pedagogues, and others. The disputes even produced the usual Soviet charges against Freudianism. These charges were quite similar to the earlier accusations of cybernetics: that Freudianism aspired to be a universal doctrine applicable not only to medicine but even to sociology, economics, and history. [77]

We have already mentioned the philosophical argument which Kolman and others presented to justify the study of thought, the brain, and the higher nervous system

with methods from the "exact" sciences, that is, that man is made up of "matter in
motion" just like everything else. Nevertheless, this question had a long life in the
USSR, and it seems to have been nearly a decade before the lingering point was
accepted that the problem was not so much man OR machine as it was man AND
machine. The objectors to "machine thought" were losing ground behind the Party's
demand for increased automation and computer technology, as well as the arguments
of spokesmen such as Kolman, Glushkov, and Berg. [78]

One of the important spokesmen to reemerge in the Soviet controversy about cybernetics was P. K. Anokhin. In 1961 he asserted:

If we accept the most widely prevalent opinion that cybernetics is now the most
synthetic and at the same time the youngest scientific discipline, which is still out-
growing the stage of its infancy, then one may make so bold as to state that physiology was the first solicitous nurse of this infant. [79]

Although this Soviet priority existed, Anokhin himself does not seem to have made
a strong effort to exploit it or to present himself as a "Soviet father" of cybernetics.
His position has consistently been for increasing independence in scientific research,
and for the position that all possible problems were not solved by either Pavlov or
cybernetics. His arguments have been consistent with the demand to use the best
of world and Soviet science and technology. [80] As a supportive commentary, on July
8, 1960 the journal Medical Worker provided an enthusiastic account of Wiener's
Moscow visit during the International Conference on Automatic Control. It noted
with obvious satisfaction Wiener's praise of Anokhin's work as well as general Soviet
progress in automation. On the matter of "new directions" it quoted Wiener:

Although cybernetics has as yet failed to pose or solve a single new biological problem, it has opened a new approach to the solution of a number of problems and to
the discovery of connections between phenomena which, heretofore, had been left
unnoticed.

V. V. Parin also came to the forefront of this controversy with his article, "The
Authority of Facts—Concerning the Scientific Heritage and Dogmatism." [81] Parin
observed that in any period of scientific growth there were bound to be difficulties
in assimilating new facts and ideas that had to replace the old. However, this problem
was further complicated in the USSR by the effect on science of the Party's elimination
of the Stalin cult and the restoration of "democratic, Leninist norms." He unequivocally
stated that the main harm endured during the cult was the proclamation of a single
viewpoint as the supreme truth. Parin gave as his first examples Stalin's Economic
Problems of Socialism in the USSR and the grassland rotation schemes in Soviet
agriculture. The problem was one of turning worthy scientific heritage into dogma,
substituting quotations from the masters for research.

Parin devoted some space to the fact that a collection on cybernetics that had been prepared by the USSR Academy of Sciences in 1961 drew forth critical reviews and reservations to "cybernetics and biology." Parin argued that the development of new methods from the "exact sciences" for research in biology was not only necessary, but called for in the Party Program. In the same way Parin described an incident from the Literary Gazette in 1961 which involved P. K. Anokhin. In response to an article by Anokhin, an (unnamed) "learned physiologist" wrote a letter condemning Anokhin for his ignorance of physiology and philosophy, demanding that he be "exposed" for departing from classical Pavlovian physiology (citing the "Pavlov Sessions" in support).

Parin pointed out that Pavlov himself had supported independent schools in physiology, including those of Orbeli, Anokhin, and Bykov. Later, however, scientific disputes degenerated to the use of out-of-context quotations, with "administrative pressure sometimes used as the main argument." Most of the article is a heated condemnation of Bykov's monopolization of Soviet physiology, a time when facts were changed to fit the "only correct" theory, when physicians had to perform on an ad hoc basis in the absence of a really correct theory, and when research consisted of singing paeans to the monopolists who regarded research meritorious to the extent that it was unoriginal.

Parin linked cybernetics and new research in physiology with the Soviet space program. Had the "dogmatists' veto" been truly effective, he argued, then important new laws of physiology would not have been discovered and Soviet successes in space medicine and biology would have been impossible.[82]

With time, the ties between cybernetics and physiology helped produce independent research in physiology, medicine, and pedagogy.[83] This evidently arduous process in physiology shared some features with the expansion of cybernetics into other areas, that is, the association of old figures and earlier research which had been eclipsed with new problems and independent research. Although the monopolists' hold was weakening in some of the fields, the advocates of independence made every attempt to avoid denigrating the old heroes (for example, Pavlov or Michurin) and cautiously avoided going so far as to pose the threat of "revisionism." Every effort was made to link the new directions to priority research in pressing problems "raised by life itself," and to place these endeavors at the service of the Party leadership.

The final area of the expansion of cybernetics with which we will deal is the social sciences, particularly law and economics. In some ways this was the most complex development, one even more influenced by Stalinist history than the natural and life sciences. The reasons for this were simply that research activities in these areas were directly related to the administration and organization of the state, involving the role and function of the Party and its leadership. These were the areas in which the expressions of confidence in science and research which had been made by the post-Stalin leadership were put to their severest test. The strength of that faith and confidence, as well as the "pragmatic utilitarianism" of the leadership, often seems to have been belied by an absence of achievements of implementation. This was especially noticeable at the All-Union level, which is where these "global conflicts" tended to be fought out.

One of the areas in which cybernetics seemed to be making noisy inroads was Soviet law. In this case, there were a variety of themes developed which were related to the de-Stalinization and received great attention in discussions far removed from cybernetics. Three such prominent directions were (1) the use of statistical and sociological methods to study the source and extent of crime in the USSR; (2) the use of computers and formal logic to obtain a consistent, noncontradictory and unambiguous body of Soviet laws; and (3) the use of quantitative methods (for example, "pattern recognition") to obtain more precise rules of evidence in order to guard against the "intuition" of the court.[84]

The pursuit of the sociology of crime was on delicate ground for several reasons. First, the Soviets have generally been quite reluctant to provide any data concerning the magnitude and nature of criminal activity. Second, to deal seriously with the problem would require the abandonment of the "voluntarist" formula that crime in a socialist state can be due only to remnants of "bourgeois mentality" in the system. Third, such research would have to be divorced from similar studies in the West for ideological reasons. Thus, for example, an article "On a Reference-Information Service in the Field of Law" had to include

Recently the American press, with its characteristic sensationalism, reported on the "research" work carried out by the couple S. and E. Gluck... who "having devoted the larger part of their lives" to the collection of data on 2500 criminals, concluded that it is possible to reveal the roots of crime with the aid of a computer. At the same time, the roots of crime in the USA are well known without any computers.[85]

Nonetheless, the author assures his reader that without reducing social science to technology, one can apply mathematical logic, game theory, operations research, queuing theory, and so forth to problems of optimum control in the USSR.

The quest for "scientifically substantiated" rules of evidence was closely related to the simultaneous campaign against the excesses of "socialist legality" which had occurred during Stalinism. In particular, the discussions were directed against A. Vyshinsky and his "subcult," in which rigorous proof in the courtroom was allegedly

denied in favor of the prosecutor's "intuition," "maximum credibility," and a "live
sense of truth." Criticisms of the arbitrariness in legal practice were current with
the Twentieth Congress. These criticisms were echoed in several papers from the
Kudryavtsev collection:

Avoiding the mention of intuition, certain authors replaced this concept with a "live
sense of truth" (A. Y. Vyshinsky) and other definitions which, like the idealistic
understanding of intuition, appear to be beyond logical analysis and precise description.[86]

What did all this have to do with "cybernetics"? There are several answers. On
the one hand, there were technical attempts to devise more quantitative rules of
evidence and systematic analysis of Soviet crime. Second, it was clear that there
were Soviet lawyers and scientists who were willing to combine their interests and
carry on old battles under a new cover. Finally, in a variation of the "kto-kovo"
question (which we might phrase as "who was using whom"), it is likely that scientists
and lawyers used each other for their own tactical advances. Perhaps we may indicate
the seriousness of their involvement by noting that six years elapsed between the
establishment of the Cybernetics and Law Division of the Scientific Council on Cybernetics and its First All-Union Co-ordinating Conference in 1965.[87]

The final area of interest in the expansion was "economic cybernetics." This area
was one of those in which the circular faith in the mutual support of cybernetics and
socialism was strongest. Phrased differently, the argument was that cybernetics
could achieve optimality, coherence, and rationality in the administration and organization of a system whose optimality, coherence, and rationality guaranteed the maximum use of cybernetics. Again, the connections between cybernetics and economics
were sometimes made only through computer use and quantitative methods.

The emergence of cybernetics and economics occurred mostly in three areas that
came to attract renewed Soviet interest in the discipline of economics. These were
in plan construction, plan implementation, and the development of an information network adequate to sustain the first two activities on a continuous basis. The first two
problems were intimately connected to the rebirth of a school of mathematical economics. Such studies had fallen into disuse during Stalin's rule but began to attract
interest again in the middle fifties. Topics concerned price formation, optimum
planning, centralization of control, and the use of incentives (as opposed to coercion)
in running the system. The third area brought forth wide discussions about the grandiose statewide computer system for control and monitoring of the national economy.[88]

There is some evidence that cybernetics was a "Trojan horse" in the discussions
about pursuing mathematical economics and econometrics (renamed "planometrics"

by Nemchinov) in the USSR. Early discussions in the field (for example, "input-output" analysis) started in the mid-fifties at a time when cybernetics was still controversial. In 1958 Lange's book on input-output analysis was translated into Russian, as was W. Leontief's Studies in the Structure of the American Economy. Kantorovich's 1939 paper on linear programming (later developed independently by Dantzig in the United States) was republished in Nemchinov's 1960 collection, The Application of Mathematics in Economic Investigation. In 1959, Kantorovich published his book, The Economic Calculation of the Optimum Use of Resources. It was evidently not until after cybernetics had been legitimized that the association with economic controversies arose. [89]

Once the open discussions were under way, there was the familiar tendency for them to degenerate into name-calling sessions posing "bourgeois" against "dogmatist."[90] Part of the reason for this was that, ironically, linear programming and input-output analysis surfaced in the Soviet Union as devices from exclusively "bourgeois economics."[91] However, there were three questions that were posed in a serious way for the discussions: Should there be a separate "school established between economics and mathematics?" If such a school existed, should it have an impact on planning and administering the overall system (as opposed to individual enterprises)? What were the social, political, and economic implications of the techniques used in optimization?

By 1964 the first question was resolved, and the Central Economics-Mathematics Institute was established as an Academy Institute with N. P. Fedorenko as its head. The journal Methods of Mathematical Economics quickly appeared. The second and third questions have not been answered, although a good deal of work has been done (and, apparently, applied) in suboptimization.[92] Questions of optimization continued to be vigorously discussed, involving issues of the proper criterion of optimization, questions of price formation, and the political consequences of the bases of theories of optimization. Spokesmen for mathematical economics have taken great pains to assign the task of selecting criteria of optimization explicitly to the Party and state leadership. Even on this task, the leadership had evidently not seen fit to commit itself with a clear decision of All-Union scope. [93]

Problems of price formation were extremely difficult, dealing with both ideological and economic questions (for example, the labor theory of value). They were also related to implementation of the plan, and a few suggested the use of economic incentives and price formation to replace the older "administrative methods." [94] Other major questions came forth, including the problem of balancing central control with local autonomy and enterprise initiative. One of the most embittered disputes centered on the use of Kantorovich's "objectively conditioned indicators."[95] These questions have still not been settled, and in 1965 the reader of the Economics Gazette (March 10) encountered a series of letters under the title "Pro and Contra." Kantorovich, Nemchinov, and Novozhilov were the only economists nominated for the Lenin Prize that year, receiving it only after great difficulty. The purpose of the "Pro and Contra" series was to try to decide whether the three were pillars of Soviet economic progress or bourgeois heretics trying to introduce "the anarchy of the bourgeois market place" into the Soviet system.

A major problem that emerged in connection with the unified computer system was also the proper degree of centralization. In this discussion, the supporters of the computer system took great pains to deny that they were trying to interfere with enterprise initiative, and to assert that the economy neither could nor should be planned "to the last bolt."[96] Another key question is who should have the responsibility for designing, supplying, and running the system. Given the jurisdictional disputes and questions of whether such a network should be organized by ministry or by territory, little had evidently been done to decide on such issues as the basic structure and function of the network itself. Glushkov had argued that one should go ahead and build the centers while the disputes go on, since they can be linked later, and in the meantime the computer capacity of the country would increase (in line, by the way, with one of Glushkov's major interests). Some work has been done in building the centers, but after considerably more than a decade, it appears that the computer network has been severely hampered by the very problems it was supposed to solve— confused and inconsistent reporting procedures, jurisdictional battles, and bureaucratic inertia, inefficiency, and red tape.[97]

In the meantime, questions of "economic cybernetics" seem to have lost the center of attention, first to mathematical economics and the unified computer system per se and then to the rush for "economic accountability" that quickly followed Khrushchev's ouster. During that time, however, several independent directions and schools of economics have won recognition, even if the implementation problem has been severe. The fact that the initial ties between "cybernetics" and economics may have been largely unessential technically may have been the source of Birman's gentle chiding of those who still used outdated formulas (which Birman himself had used often).

Incidentally, it is time to stop talking about "the application of mathematicial methods and electronic technology."...There is no need at all to distinguish the use of mathematics and electronics from other means of solving economic problems, or, even more important, from the same correct economic position as any other, and the use of mathematics and electronics is a purely technical question.[98]

SUMMARY AND CONCLUSIONS

It would seem that the Soviet love affair with cybernetics has clearly passed its peak, and in fact it did so shortly after Khrushchev's ouster. This is not to say that serious scientific and technical work in fields once covered by the aegis of cybernetics has stopped, but rather that the work is pursued more quietly on a less grandiose scale in "independent directions" that were rejuvenated during Khrushchev's years. Current popular articles on cybernetics are usually more technically specific and less polemical in tone. The days seem to have passed when cybernetics and new directions in science had to be defended automatically from the attacks of orthodox spokesmen (although now they must evidently assure their profitability to the state). The days also seem to have passed when cybernetics offered the polemical resources of popularity and official approval for the advocates of new approaches and methods in diverse areas of Soviet science.

There are several possible reasons for the decline in Soviet preoccupation with cybernetics as a political and polemical symbol. One may be that many of its more vocal enthusiasts largely achieved what they set out to do: win acceptance for independent research in previously moribund fields. Another is simply that the term cybernetics became so warped by its use in the various battles that have been described, and so inflated from its use by technically irresponsible interests and people who wanted to share the popular light of cybernetics, that the term collapsed into meaninglessness. After diverse new directions found institutional and material support for their researches, it seems that it was not longer desirable to be constrained as a "subfield" by technically unessential links to cybernetics.

Another possible reason for the current lack of fervor about cybernetics is simply that some Soviet spokesmen for the science promised miracles in some areas: "artificial intelligence," machine translation of languages, or "made-to-order" plant and animal species that rivaled the claims of Lysenko. These problems may have simply turned out to be more difficult than initially estimated, and a bit of temperance began to be injected into the discussions by selected spokesmen for cybernetics.[99] There are probably technically responsible figures in Soviet science and technology who would prefer to be dissociated from the liabilities of this legacy, and perhaps for this reason there is a more frequent use of the term "control theory and management."

Of greater interest is the fact that people who have been among the foremost technical spokesmen for cybernetics have started to espouse the cause of "large systems theory." For example, in 1965 Glushkov wrote on the great difficulty in heuristics for chess-playing and language-translating machines:

The nature of all these difficulties ultimately turns out to be the same thing. The problem is that from the example of modeling the problem of translating from one language to another and other analogous problems, we encounter a new class of problems which are brought together in the so-called large system theory. A precise definition of the subject of large system theory which would be considered generally recognized has not been worked out as yet. But the heart of the matter here is in finding methods for devising complex control systems which in principle cannot be made simple due to the specifics of the problem being solved.[100]

Another example of this shift comes from D. Gvishiani, who expressed his interests in "systems analysis" and "scientific management" in 1966.[101] He noted as the motive for Soviet interests in new management techniques the continued lag (by a factor of 2) behind U.S. labor productivity, and posited the familar demand both to learn from the West and guard against the importation of methods that might be inimical to Soviet goals. He also expressed his approval of Glushkov's rather mild definition of cybernetics as "the science of the general laws of handling information in complex control systems....Cybernetics is no more than a concrete method of inquiry."[102] Gvishiani continued:

At present, the task of applying cybernetic principles to control through the use of electronic computers is being rather extensively introduced in our country. Unfortunately, the practical results are still absolutely unsatisfactory....

In view of these facts, it is very important to take into account a multitude of very diverse factors and to examine quite extensively the solution of problems in the different sciences. This is done by so-called systems analysis, which is the study of designing large-scale organizational and technological systems.

The development of studies in systems analysis reveals another line of awareness in Soviet thought about "cybernetics" and its effective utilization. In this connection Glushkov, for example, has long been interested in a "systems approach" to computer development, that is, the notion that someone should actually be held responsible for providing the means of effectively and optimally using the computer ("software") and such mundane matters as actually installing the thing at the customer's site and maintaining it. This is only one aspect of the total "systems awareness" that exists in some parts of the Soviet Union, an awareness which is beginning to reveal that the sheer number and types of "cybernetics devices" produced may be a poor indicator of productivity unless one can also assure the effectiveness and purposefulness of their utilization.[103]

There seems to be a widespread (although not universal) opinion in both the United States and the USSR that the final test of scientific success and excellence is determined by "life and practice," that is, the practical payoffs of science in making the lives of the citizens more productive and/or easier to live. This brings into consideration the

problem of implementation, a problem that has long proved very difficult for the
Soviets. It is the area where Soviet bluster must be separated from Soviet realities.
The Soviets themselves have devoted an enormous amount of attention to the topic
of implementation, using everything from massive reorganization of control to the
use of economic incentives to speed the use of scientific findings. [104]

There are three aspects of the implementation problem which merit discussion
here. The first depends upon the characteristics of the supporters of cybernetics in
the USSR. The second involves the political problem of trying to satisfy diverse interests with at best rather simple decisions that are to be applied through the length
and breadth of the country. The third concerns the difficulty of finding reforms that
would not undermine the power and influence of the reformers themselves.

We have identified one source of support for cybernetics in the USSR as coming
from those who wished to use the issue as a way to renovate some "schools" and
"independent lines" of investigation that had been suppressed under Stalinism. We
have also seen that the cybernetics controversy was only one way by which these
people sought the right and support to ask new (and old) questions in their fields. This
right and support was often gained in the Khrushchev years—no easy achievement—
but there are two things to note about the victories. First, in these fields there was
often a great deal of catching up to do, and although Soviet scientists returned to their
laboratories it is not so evident that they have yet emerged with much to be implemented. Second, even where the scientists have won the right to do research, it is
evident that in some fields they have not really won much in the way of capital, personnel, and equipment. Workers in fields of biology, physiology, and medicine have
gained rights that have a hollow ring in view of the fact that basic Soviet priorities in
industrial and military science and technology have remained largely unaffected.

Another group of people who supported cybernetics in the Soviet Union were those
who wished to pursue the development of new topics in science, such as computers,
heuristics, and so forth. Many of these problems are harder than they first seemed,
and while greater sophistication may have come forth in terms of an appreciation of
the magnitude of the problems, practical successes have probably not matched the
earlier promises. In some fields, such as social, economic, and technological forecasting, or "scientific management," the early assurances of quick results are only
now being followed up with serious attempts at systematic study.

The second aspect of the implementation problem concerns the difficulty of finding
a single policy or reform to be implemented which would satisfy the exceedingly diverse interests of various segments of Soviet society. This problem seems to be even
more severe in the USSR due to something like Soviet omnipotence fantasies that stress

the payoffs of the single policy applied uniformly to all aspects of Soviet life.[105]
Coupled to this is a seeming inability to differentiate (by relevance and importance)
the experience from diverse sectors of Soviet society. There often seems to be a
kind of delight at confronting expert testimony with counterevidence from "production
worker," with both claim and counterclaim being treated equally. Such procedures
rarely go so far as to result in a kind of interest aggregation and articulation from
Soviet society.[106]

Even the discussions about cybernetics within single scientific disciplines were
rather complex. The "lowest common denominator" was that cybernetics could serve
as a shield behind which old investigations and arguments could be renewed. Even
here the issues varied sharply from field to field (for example, centralism in economics, "machine thought" in physiology and psychology, classical genetics in biology, the rules of evidence in law). This complexity within single fields becomes
still more difficult to analyze when interdisciplinary interests are considered (for
example, the attitudes of biologists toward physicists, of physiologists toward psychologists and cyberneticians, of physicians toward engineers, of economists toward
mathematicians). Finally, to this must be added the interests of groups outside the
disciplines, in particular party and government officials and industrial managers.

There are two points to this observation of the complexity and interests that were
expressed in the cybernetics affair. First, one simply could not get very far into an
analysis on the basis of a simple model (for example, the "two person zero-sum
game") in which the party is counterposed to scientists, or liberals to conservatives,
in which what one person gains the other loses. The second point is that some Soviets
have themselves acted as if this was the only model that pertained. On this basis they
counterposed "bourgeois science" to "socialist science," and to some degree this
tendency persists among some of the foremost spokesmen for cybernetics (for example,
Berg and Glushkov). One aspect of this is the persistence of the demand to create
"Marxist-Leninist bases" for fields associated with cybernetics, not "empirical-
theoretical bases."[107] At the more polemical level, there is the continued tendency
to condemn one's scientific opponent by associating him with an allegedly like-thinking
figure from "bourgeois science."

The third and final aspect of the implementation problem concerns a dilemma inherent in many reform movements. It involves the fact that programs must be worked
out which not only satisfy a variety of (often contradictory) needs, but they must also
be undertaken in a way which does not undermine the legitimacy and status of the
reformer himself.

It seems that the areas most affected by the development of cybernetics in the USSR have been those that had the least direct connection with the role and structure of the Communist Party and Soviet government. In those areas where there might have been a more direct impact on the political system (for example, economics, sociology, law, or even ideology) the impact of cybernetics seems most shallow and the problems of implementation most severe. The grandiose plan for the national system of computer centers for directing the national economy was perhaps the closest link between science and Soviet state politics, and the system has simply not progressed very far owing both to political and technical indecision and battles over centralization and jurisdiction. This must be particularly evident to the Soviets themselves, not only because of renewed insistence on implementation and practical payoffs from science and technology.

In the meantime, we can better understand the position of Academician Berg as well as other, unnamed, supporters and detractors of cybernetics.

Meanwhile, it is not now possible to recognize the position (of cybernetics) as completely secure. Some who by nature are apt to be carried away are inclined to exaggerate somewhat the potential of modern cybernetics, which is not especially dangerous because life itself will introduce its corrections....

The main obstacle in the way of the wide use of the possibilities of cybernetics is the protection of the old from the new, this camouflage of its own conservatism by slogans of skepticism and the "caution"; "Look what didn't happen".... [108]

NOTES

1. Only a very few rather recent works have addressed the question as to whether there may be some positive and negative features of automation that are common to socialist and capitalist systems. For example, V. V. Alekhin ("Experience in Concrete Investigations of Man's Place under Conditions of Automated Production," Herald of Moscow University, Philosophy, No. 4, 1967, pp. 53-63) provided similar accounts of tedious assembly-line work at both a U.S. automobile plant and the Likhachev automobile plant in the Soviet Union. Noting the argument that the consequences of the scientific-technological revolution are exclusively good in socialism and exclusively bad in communism, he emphasizes: 'It seems to us that the contemporary scientific-technological revolution has its own relatively independent characteristics, which operate under both socialism and capitalism, exerting a profound transforming influence on the life of contemporary society." He then concludes that the dreariness of labor in the Soviet plant is due to the character of assembly-line work; the lack of sufficient specialists; only those processes are automated that contribute directly to enterprise profit, his article is translated in the RAND publication Soviet Cybernetics: Recent News Items, Vol. 2, November 1968.

2. Even Wiener entitled the second volume of his autobiographical works I Am a Mathematician.

3. On many occasions, it was apparently thought sufficient merely to mention a connection with computers or even "information" to establish the tie with "cybernetics." For example, A. I. Berg, chairman of the USSR Academy of Sciences Scientific Council on Cybernetics, mentioned that "there are 73,000,000 people engaged in so-called informational activity in the USSR" ("Problems of Information," Trud, December 17, 1968). From the context, it appears that he meant the number of teachers and students in the USSR, and he was merely broadening the base of the bandwagon for one of his current pet projects: the quest for efficiency and effectiveness in the use of teaching machines and programmed instruction (on occasion, referred to as "pedagogical cybernetics" in the USSR).

4. For example, the article with which many Soviet and Western reports of cybernetics begin was "Whom Cybernetics Serves," published under the pseudonym "Materialist" in Questions of Philosophy, No. 5, 1953 (hereafter this journal is referred to as V.f.). The article is included in the section of the journal entitled "Criticisms of Bourgeois Ideology"; this article has been translated and published in the RAND series Soviet Cybernetics: Recent News Items (hereafter SC:RNI), Vol. 1, July 1967, pp. 27-49.

5. For example, the problems of "artificial intelligence" ("thinking machines, heuristic programming, machine translation of language, and so on) seem to be much harder than some early enthusiasts anticipated, both in the USSR and the United States. See Hubert L. Dreyfus, "Alchemy and Artificial Intelligence," RAND publication P-3244, December 1965. If my understanding is correct, the Soviets planned to translate and publish this excellent but controversial article in the USSR. To my knowledge, it has not yet appeared there.

6. Some of these priorities have since been substantiated. The earlier lack of recognition in the West for these discoveries has been partially attributed to the language barrier and Russian secretiveness.

7. The campaign to mix ideology, philosophy, and nationalism was not new to either the Zhdanovshchina or the Soviet system. An engaging study of science under the tsars is presented in A. Vuchinich's Science in Russian Culture to 1860 (Stanford, Calif.: Stanford University Press, 1963). The so-called "Deborinite-Mechanicist Debate" of the late twenties and early thirties (which had many similarities to the central arguments about cybernetics) is excellently covered by D. Joravsky, Soviet Marxism and Natural Science (New York: Columbia University Press, 1961).

8. A. Vuchinich, The Soviet Academy of Sciences (Stanford, Calif.: Stanford University Press, 1956), pp. 105-106.

9. For example, there was a bitter exchange between the Soviet physicist V. A. Fok and the physics editor of Questions of Philosophy, A. A. Maximov (see Maximov's article in Red Fleet, June 13, 1952, and the articles by Fok and Maximov in V.f., No. 1, 1953, pp. 168-194). Fok was accused of concealing his defense of "reactionary philosophical theories in quantum mechanics" with statements about the usefulness of dialectical materialism.

10. The links in the life sciences begin to be evident in these "reorientations." For example, after the meeting on August 28, L. A. Orbeli was dismissed as the chief of the biology department of the USSR Academy of Sciences. L. A. Orbeli had been the closest worker of I. P. Pavlov, and the attack on him at the "Pavlov Sessions" of the USSR Academy of Sciences in 1950 make this affair seem in some ways like a small "succession crisis" in Soviet physiology.

11. T. D. Lysenko, Reports of the Lenin Academy of Agricultural Science, No. 10, 1951, pp. 2-5.

12. Pravda, June 29, 1950, pp. 3-4. Translated in the Current Digest of the Soviet Press (hereafter, CDSP) Vol. 2, No. 25, pp. 3-9, and No. 27, pp. 8-12.

13. See, for example, the article by the president of the USSR Academy of Medical Sciences, N. N. Anichkov, in Priroda (Nature), Vol. 42, No. 2, 1953, p. 4.

14. In his Twenty Years of Experiments (Moscow, 1938, p. 275) Pavlov cautiously speculated that conditioned (learned) reflexes of the parent might become uncon- ditioned reflexes in the offspring if conditioned reflexes were based on unconditioned reflexes. Writing in Priroda in 1952 (Vol. 41, No. 7, pp. 3-16), Bykov claimed that this was exactly the mechanism which pertained, so that Pavlov's teaching not only refuted Wiesmann's theory but provided a "true natural scientific method for the transformation of the animal organism in the interest of man" in accordance with the doctrine of Michurin.

15. Indeed, some of this work was established on the basis of diligent laboratory in- vestigations. Other portions of the work, particularly that performed or supported by Lepeshinskaya and Lysenko, sometimes defied reproduction in laboratories out- side their personal control. To say the least of it, it is the capacity for universal reproduction and verification which is one of the most important differences between science and art. The virtues of systematic replication seemed to be lost on some Soviet ideologists and philosophers of science, who would apparently have to be re- minded later that Soviet atoms and American atoms and British atoms were identical (see, for example, V. Orlov, "Notes from the All-Union Conference on the Physics of High Energy Particles," Izvestia, May 20, 1956, p. 3).

16. The Soviets themselves frequently use a word that is highly appropriate in de- scribing these attempts to disguise internal battles by using universal, ideologically explosive terms: it is "obscurantism."

17. "Decree of the Scientific Session of the USSR Academy of Sciences and Academy of Medical Sciences Devoted to Problems of Academician I. P. Pavlov's Teaching in Physiology," Pravda, July 14, 1950, p. 3; CDSP Vol. 2, No. 28, pp. 31-34.

18. K. M. Bykov, Priroda, 1952, pp. 3-16.

19. The "Arakcheyev-type regime" charge played a role in debates about schools in science which went on after Stalin's death, and during this time it assumed different meanings and targets. It took its name from Count Arakcheyev (1792-1834) who was known both for his development of military establishments and the ruthlessness of their adminstration. For Stalin's "Letters," see Pravda, June 22, 1950, pp. 3-4, translated in CDSP Vol. 2, No. 21, pp. 3-9.

20. See, for example, Yu. Zhdanov, "Certain Conclusions from the Session on Physi- ology," Pravda, July 28, 1950; translated in CDSP Vol. 2, No. 30, pp. 23-26. The reference to Beritov is to the Georgian physiologist, I. S. Beritashvili.

21. For example, Lysenko's foremost antagonist in the thirties was the renowned geneticist, N. I. Vavilov, brother of the president of the USSR Academy of Sciences, S. I. Vavilov. Writing in Vernalization, No. 1, 1939, N. I. Vavilov said that he con- sidered Lysenko's doctrine to be a great achievement of world and Soviet science which

deserved a place in textbooks on genetics and selection and which, in proper application, would doubtless yield valuable results. Nevertheless, the doctrine developed by Michurin and Lysenko should not overshadow the total content of genetics and selection, and the main task should not be the replacement of modern genetics but its further elaboration and development (this is rather similar to arguments about "socialist science" in physics, where the call was often not to give up the theoretical and empirical successes of quantum theory but rather to develop it further on the basis of dialectical materialism). Vavilov was arrested in 1940, and is said to have died in a Soviet camp in 1943. A detailed account of this and other incidents in the development of "Socialist biology" is presented in Zh. A. Medvedev, The Rise and Fall of T. D. Lysenko (New York: Columbia University Press, 1969).

22. A. Berg, "Problems of Control and Cybernetics," in A. A. Ilin, V. N. Kolbanovsky, and E. Kolman, Philosophical Problems of Cybernetics (Moscow: 1961). Translated in entirety as number 11503 by the U.S. Department of Commerce in the Joint Publication Research Service (hereafter, JPRS), p. 122.

23. More detailed analyses of the philosophical issues than will be presented here are available from a variety of Western works. Perhaps the most detailed is L. R. Kerschner's unpublished Ph.D. dissertation, "Cybernetics in the Judgment of Soviet Philosophy," Georgetown University, 1964. A briefer but also excellent account is provided in M. W. Mikulak's article in C. Dechert, ed., The Social Impact of Cybernetics (Notre Dame, Ind.: University of Notre Dame Press, 1967), pp. 129-159. C. Olgin has written on the philosophical implications of cybernetics in the Soviet Union in five articles published in the Bulletin of the Institute for the Study of the USSR: "Soviet Ideology and Cybernetics I and II," February and June, 1962; "Science, Ideology, and Cybernetics in the USSR," July 1966; "Speculative Cybernetics," August 1966; and "Cybernetics and the Political Economy of Communism," October 1966. L. R. Graham has provided two excellent articles: "Cybernetics in the Soviet Union," Soviet Survey, No. 52, July 1964; "Cybernetics," in G. Fischer, ed., Science and Ideology in Soviet Society (New York: Atherton, 1967).

24. If these elementary propositions seem merely scholastic, it is perhaps because in the West we are more likely to take their truth (or untruth) for granted. In the USSR, we find one of the leading cyberneticists (V. M. Glushkov, "Thought and Cybernetics," V.f., No. 1, 1963, pp. 36-48) using Gödel's Proof to defend simultaneously cybernetics, mathematical logic, formal linguistics, and dialectical materialism. Reasoning by analogy (which comes so easily to cyberneticists), his argument ran as follows. Gödel showed that even the arithmetic of the natural numbers cannot be formalized by a formal language with a prespecified finite number of basic facts (axioms), since one can obtain unprovable assertions that are a consequence of these axioms. We can prove these only by adding an axiom, and the new set will generate a new unprovable assertion. This, Glushkov contends, is analogous to the increase of scientific knowledge. "This constitutes a natural scientific interpretation of the following epistemological principle of dialectical materialism: there are no unknowable things in the world; any regularity which is unknown today can be discovered in the future, although at no time will this result in the attainment of absolutely complete knowledge."

25. Soviet philosophers such as A. A. Maximov were still in the throes of sorting out the matter-energy relationships that followed from special relativity. As his Soviet detractors noted, Wiener did not help much with his assertion: "Information is information, not matter or energy. No materialism which does not admit this can survive at the present day." Cybernetics, 2nd ed. (Cambridge: The M.I.T. Press, 1961), p. 132.

26. The idea that what was a dog to Pavlov could be a black box to a cybernetician came exceedingly hard in the Soviet Union. By 1965, in The Central Ideas and Philosophical Principles of Cybernetics (Moscow: Thought Publishing House), V. D. Moiseyev pointed out that the method of the "black box" was in fact much older than cybernetics, and had been invaluable to medicine in drug research and to radio engineering in diagnostic studies. What Pavlov said about the conditioned reflex was nothing other than the "black box," translated in entirety as JPRS 35136, p. 67.

27. D. A. Gushchin, "The Information Category and Certain Problems of Development," The Herald of Leningrad University, Series in Economics, Philosophy and Law, Vol. 23, No. 4, 1967, pp. 55-63; translated in JPRS 45274. Gushchin's emphasis. This vivid and generally pro-cybernetics article also included the note that "it has become clear that our electrical engineers and computer enthusiasts should either cut the chatter or accept the serious indictment that they are composing science fiction in order to tickle the reader's fancy in the pursuit of easy money and cheap popularity." In the above attempt to bolster the dialectical materialist foundations of cybernetics, the author tries to pull the rug out from under the enemies of cybernetics by asserting that not only does cybernetics not violate the divisions of Engels' typology, but cybernetics actually investigates an essential criterion by which one may determine the qualitative nature of a particular type of motion of matter. The facility with which spokesmen for the field breezily include information as a fundamental entity belies the difficulty encountered in considering it in Soviet materialism. See, for example, A. I. Berg's "Interview on Programmed Instruction," Soviet Kirgiziya, Frunze, July 26, 1966, p. 4. "It is precisely the theory of optimizing control...with the least expenditure of information, material, and energy, as well as time, effort and human labor, which is the basic goal of cybernetics." (My emphasis.)

28. These two slogans have a long history in the Soviet Union, but their use in the middle fifties entailed a shift from the targets of Zhdanovshchina. This shift carried important implications not only for cybernetics but for the "monopolists" of science as well. The charges of "Arakcheyev-like regime" and even "cult of the individual" were turned around and used in support of some independent schools that had been attacked earlier.

29. Considering science as a productive base (rather than part of the superstructure) of society was itself an important shift toward the "secularization" of Soviet science. There it was more ideologically neutral and (the Soviet argument went) scientists and philosophers should then cooperate to strip away the bourgeois ideology that had been "parasited" to the body of world scientific advance.

30. This was the defense undertaken by E. Kolman in November 1954 at the CPSU Central Committee's Academy of Social Sciences. The lecture was later published in V.f., No. 4, 1955, pp. 148-159. A translation of the article, done by Anatol Rapoport, was later published in the West: "What Is Cybernetics?," Behavioral Science, Vol. 4, No. 2, 1959, pp. 132-146.

31. A. N. Bakulev, "Important Problems in the Development of Medical Science," Pravda, January 16, 1954.

32. Kommunist, No. 5, March 1954, pp. 3-13; translated in CDSP, Vol. 6, No. 14.

33. Thus, we have Stalinist slogans (from the "Letters on Linguistics") being used against his "pet" scientists before Stalin was himself denounced. It is also important

to remember that this article was not definitive. The controversy is marked by the point that Lysenko received the Order of Lenin in February 1954. Curiously, at that time the Order was also given to A. N. Kolmogorov, S. L. Sobolev, and V. S. Nemchinov (Pravda, Feburary 13, 1954). Kolmogorov later became one of the most enthusiastic supporters of cybernetics, Sobolev defended cybernetics at the 1958 conference on philosophical problems of natural science, and Nemchinov (perhaps more than any other single person) was responsible for the development of mathematical economics and "economic cybernetics" in the USSR.

34. The theme of "practicality" cuts two ways. On the one hand, opening up areas for concrete investigation was a liberalization. On the other, limiting the investigation to concrete problems (as opposed to studies of basic Soviet goals, a Marxist analysis of Stalinism, and so forth) had a restraining effect.

35. Pravda, July 2, 1954, p. 2; translated in CDSP, Vol. 6, No. 26, pp. 15-16.

36. Appraisals of Einstein became a rough indicator of the progress of renovation in Soviet science, and changing perceptions of Western science. A. A. Maximov was fired as physics editor of Questions of Philosophy, and the new currents were evident in Einstein's obituary in Pravda of April 20, 1955. The obituary, signed by Nemchinov, Topchiyev, Ioffe, Kapitsa, Lavrentyev, Skobeltsyn, and Fok, was a tribute to the man who "V. I. Lenin considered to be one of the great remakers of natural science." The articles provided a brief account of Einstein's career, including his election as an honorary foreign member of the USSR Academy in 1926. The article also contains a bit of dialectical editorializing of use in the campaign against dogmatism, that is, the new theory demonstrated that "laws known at each stage of the development of science are not the ultimate truth, that they change with the appearance of new data." The article concludes with the assertion that although Einstein made occasional concessions to idealism in his philosophical outlook, this should in no way detract from the extraordinary significance of his scientific work.
    An exhaustive review and analysis of the Soviet problems with the theory of relativity may be found in S. Müller-Markus, Einstein und die Sowjet Philosophie (Dordrecht: D. Reidel Publishing Company, 1960).

37. The difficulty that attended the establishment of intellectual independence in some areas of science was evident even in 1955. In spite of exhortations for social scientists and others to do practical, relevant research, P. Mstislavsky and others were severely criticized for their stand in favor of a relative increase in consumer good production. It was announced that "revisionists" had tried unsuccessfully to pressure the journals Questions of Economics and Questions of Philosophy into holding discussions of "basic questions of Marxist-Leninist theory in economics." Topchiyev condemned this as disguising revisionism propositions under the guise of open discussion. See A. V. Topchiyev, "Basic Results of the Scientific Work of the USSR Academy of Science in 1954," Herald of the Academy of Sciences, No. 3, 1955; translated in CDSP, Vol. 7, No. 29, pp. 3-10. It is perhaps not entirely without irony that somewhat later V. Dyachenko, director of the Academy's Economics Institute, attributed the lack of discussion in economics to a conservative attitude on the parts of some officials who tended to avoid issues on the basis that they were "too controversial." See "On the Tasks of Research in the Field of Economics," Questions of Economics, No. 10, 1955, pp. 3-8; translated in CDSP, Vol. 7, No. 50, pp. 13-18.

38. V. Dyachenko, "Research in the Field of Economics," pp. 3-8.

39. For example, Academician L. Knunyants and L. Zubkov argued that the independent schools in physics, mathematics, and chemistry were a positive sign of progress in those fields; the existence of one school could in no way guarantee its correctness or infallibility. Pavlov himself had encouraged multiformity in science and recognized subschools headed by L. A. Orbeli, K. M. Bykov, and others. The authors asked for an end to labeling opponents as "reactionaries" and for the recognition that the ultimate truth in science is not necessarily in the immediate possession of only one of the disputants. See their "Schools in Science," Literary Gazette, January 11, 1955; translated in CDSP, Vol. 7, No. 2, p. 7.

40. A. V. Topchiyev, "Basic Results of Scientific Work," pp. 3-10.

41. Here there were differences by field of science. In areas where the social impact of technological decisions seemed more tangible and likely to involve the party and control of Soviet society, the disputes remained viable for a long time. Economics and sociology are thus sharply constrained in the freedom of investigation, even where ties were made to "cybernetics" or fashionable uses of computers and quantitative techniques.

42. Of course, an easy way out of the impasse was to blame the entire affair on the "bourgeois press" and the gullibility of some Soviets who believed it. At a more polemical level, the charge could then be made that those who resisted the new trends were "objectively" aligned with the forces of reaction. For example, in April of 1955, S. L. Sobolev, A. I. Kitov, and A.A. Lyapunov charged that "several ignorant bourgeois journalists promoted publicity and cheap speculation about cybernetics," and Soviets who objected to cybernetics really did not know anything about the subject except what the bourgeois press said.

43. In a similar way, I. B. Novik in his book Cybernetics: Philosophical and Sociological Problems (Moscow: 1963) described Wiener as an "unconscious dialectician."

44. Kommunist, No. 5, 1955, pp. 10-14; translated in CDSP, Vol. 7, No. 32, pp. 7-9.

45. For example, see "For the Leninist Principal of Adherence to Party Spirit in Ideological Work," V.f. editorial, No. 6, 1956; translated in CDSP, Vol. 9, No. 3, pp. 7-9.

46. "The Cooperation of Scientists of the Union Republics," Pravda, February 27, 1957.

47. A. V. Topchiyev, "Science and the Building of Communism," Kommunist, No. 13, September 1957, pp. 78-84; translated in CDSP, Vol. 9, No. 41, pp. 3-8. After strongly condemning monopoly and dogmatism, Topchiyev observed that all Soviet scientists noted "With deep satisfaction and warm approval" the ouster of Malenkov, Kaganovich, Molotov, and Shepilov as an example of the unceasing struggle with ideological distortions, confused dogmas, and ideas removed from life: the cornerstone of scientific work.

48. It was a busy week for Soviet scientists. On October 29 Pravda announced that B. Pasternak had been expelled from the Writer's Union. In the same issue it was the task of Academicians I. Kurchatov, N. Semenov, A. Topchiyev, A. Alexandrov, A. Ioffe, V. Fok, and Corresponding Member B. Vul to explain why the 1958 Nobel prize to Cherenkov, Tamm, and Frank was a fitting tribute to Soviet physics, whereas in literature the prize was at best tendentious and at worst politically motivated.

49. This series has now progressed to BESM-6. A. N. Nechaev, The Arrangement and Work of the Electronic Computing Machine (Moscow: 1964); SC:RNI, No. 11, December 1967, p. 76.

50. Pravda, Feburary 19, 1956, pp. 3-4. The early fifties was a period of intense Soviet interest in applied mathematics, and in 1954 Nesmeyanov spoke of the applied mathematics department of the V. A. Steklov Mathematics Institute. No doubt computer development was related to these interests rather than "cybernetics" per se. A lingering point, however, is simply that the cybernetics dispute did not halt this work. A. A. Lyapunov, in the first issue of the journal Problems of Cybernetics provides some insight into Soviet research in times of controversy: not only had the disputes inhibited publication in the field, but (more seriously) research had been inhibited. Thus, in the USSR the publication of research and the performance of research are two quite different considerations.

51. See A. N. Nesmeyanov, "Elections to the USSR Academy of Sciences," Pravda, October 11, 1953, p. 2, and A. V. Topchiyev, "On the Elections to the USSR Academy of Sciences," Izvestia, July 23, 1953.

52. A. V. Nesmeyanov, "Looking into the Future of Science," Pravda, December 31, 1955. Nesmeyanov had long taken stands in favor of interdisciplinary research, often in terms of biophysics and biochemistry. He was often reserved about cybernetics, however, sometimes treating it as merely a part of the older field of automation and remote control.

53. A. A. Bulganin, "Directives of the Twentieth Congress of the Communist Party of the Soviet Union on the Sixth Five-year Plan for the Development of the USSR National Economy," Pravda, February 22, 1956, pp. 1-6.

54. This defense was made before the CPSU Central Committee's Academy of Social Sciences, later published in V.f., No. 4, 1955.

55. This, of course, was a central argument in the above-mentioned article by Sobolev, Lyapunov, and Kitov which appeared at the same time.

56. E. Kolman, Cybernetics (Moscow: 1956); translated in JPRS 5002.

57. Their presentation, "Cybernetics and Natural Science," was published in the Proceedings of the All-Union Conference on Philosophical Problems of Natural Science (Moscow: 1959); it is translated in JPRS 7233.

58. For example, "even now the fruitfulness of cybernetics is beyond doubt in the study of ... various very essential aspects of economic processes. The material basis of this application of cyberntic methods exists in the form of electronic computers .... Therefore, one who speaks against the application of cybernetic methods to economics is also objectively against the utilization of computers in this most important field." V. D. Belkin, "Cybernetics and the Social Sciences," in A. Berg, ed., Cybernetics at the Service of Communism (Moscow: 1961); translated as JPRS 23554.

59. Lyapunov's interests in mathematics and biology evidently predated the 1958 conference. For example, in the criticisms of Lysenko that appeared in the Botanical Journal, A. L. Takhtadzhan thanked A. A. Lyapunov (and many others) for reading his article "Direct Adaptation or Natural Selection," No. 4, April 1957, p. 609.

60. See "Documents I," Daedalus, Vol. 89, Summer 1960, p. 641. These discussions saw G. Platonov (later charged with dogmatism in biology) complain that the conference gave too little attention to idealism and science, also criticizing Kolman for his uncritical introduction to Wiener's Cybernetics (translated into Russian in 1958). Professor Kostrikova attacked the geneticist N. P. Dubinin and criticized Sobolev and Lyapunov for presenting generations as a "stream of information." P. K. Anokhin spoke of the mutual assistance which cybernetics and physiology could provide each other.

61. Pravda, January 28, 1959. Khrushchev said there was a lag of a factor of 2 to 2 1/2 in industry and 3 in agriculture.

62. Pravda, October 28, 1959.

63. The underlined portions were added to the draft. The final sentence replaced the original: "It is necessary to organize the wide use of cybernetics and electronic computers and control systems in production, scientific work, designing and planning, and in accounting, statistics, and management. See CDSP, Vol. 13, No. 46.

64. Khurshchev's comments were in Pravda, April 18, 1958. Kolman's were in his article "The Philosophical and Social Problems of Cybernetics" in Ilin, et al., Philosophical Problems of Cybernetics, which, like Cybernetics at the Service of Communism, appeared in 1961.

65. Berg, "Problems of Control and Cybernetics," in Ilin, Philosophical Problems. This was probably a reference to the "statewide computer network" for planning and control of the economy which was discussed openly a few years later. See note 88.

66. "Cybernetics," A. I. Berg et al., Philosophical Encyclopedia, Vol. 2, (Moscow: 1964), pp. 405 ff. The Soviets were later involved in a "scientific organization of labor" movement (NOT) in which cybernetics played an occasional, rather amorphous part. The Soviets opened their first school of management for high-level officials in January 1971.

67. "Seek, Dispute, Dare," Izvestia, September 4, 1961. Blokhintsev, then director of the Joint Nuclear Research Institute, had earlier connected cybernetics and nuclear research via the instrumentation for the Institute's fast nuclear reactor, Pravda, May 18, 1959. Similar activities seemed to have attended the designing (and, perhaps, the selling) of the Soviet's largest accelerator, which from time to time has been referred to as the "Cybernetics Accelerator."

68. Three of the most notable were P. K. Anokhin, V. V. Parin, and A. A. Vishnevskii. Vishnevskii, head of the Institute of Surgery named for his father, now has a "Laboratory of Cybernetics" at the Institute that works on problems of medical engineering. Before 1957 (at least) the head of his Institute's physiology laboratory was P. K. Anokhin, who was pummeled by Bykov at the "Pavlov Sessions" and whose concept of "reverse afferentiation" has been hailed by some Soviets as the predecessor of the "feedback" concept. V. V. Parin has been active in space medicine, and on occasion has linked cybernetics to the Soviet space program via physiological monitoring, and so on.

69. See, for example, V. V. Parin, "Mathematics and Life," Priroda, No. 7, 1962, pp. 22-27, and his more militaristic exposition in Morskoy sbornik, No. 9, 1965.

70. Pravda, April 4 and 6, 1963, and Izvestia, April 4 and 8, 1963.

71. N. M. Sisakyan, "Biology and Cybernetics," Priroda, July 7, 1962, pp. 13-15; this collection of articles from Nature has been translated in JPRS 15284.

72. V. V. Parin, "Mathematics and Life."

73. Izvestia, October 1, 1964.

74. G. Gagarin, "Crises in the Field of Biology," Studies of the Soviet Union, Vol. 5, No. 3, 1965. This sounds like one of Parin's "icebergs."

75. Izvestia, January 22, 1967.

76. In his Cybernetics at the Service of Communism, Berg claimed that more than 5,000 articles appeared in 1961 on the application of mathematics, electronics, and cybernetics to biology and medicine.

77. See the report on the 1959 Scientific Conference on the Ideological Struggle with Modern Freudianism, P. P. Bondarenko and M. Kh. Ravinovitch, V.f., No. 2, 1959. Wiener was condemned in this article for allegedly contending that the main tenets of Freudianism were consistent with the laws of physics. P. K. Anokhin was in the opposition, asking for a more thorough knowledge of Freudianism. In 1962 P. Kapitsa was attacked in Kommunism (No. 8, pp. 60-62) for his "reductionist" assertion that just as advances in biology should be based on studies of the higher nervous activity which were founded by Pavlov and Freud (Science and Life, No. 3, 1962).

78. One tactic was to assert the social bases of thought, so that machines could never become smarter than society, although in some areas they could already out-perform man. A more simplistic argument was simply that man-made computers, not vice versa (Glushkov, V.f., No. 1, 1963, pp. 36-48). A more telling argument was that, just as the total number of power plants determines the power capacity of a country, the total number of computers determines the "informational-intellectual power" of a modern industrial nation. See Glushkov's article in Berg and Kolman, eds., The Possible and Impossible in Cybernetics (Moscow: 1963), pp. 198-204.

79. Ilin et al., Philosophical Problems, Ch. 8. Anokhin pointed out that the concept of "feedback" had existed in Soviet physiology as early as 1935. References to Ano-khins's work are quite common in early Soviet histories of cybernetics. S. N. Braines ("Neurocybernetics," in A. Berg, ed., Cybernetics at the Service of Communism) singled out Wiener, Anokhin, V. V. Parin, and E. Kolman for the "special attention" they gave to problems of control in living systems.

80. An editorial in Medical Worker, April 5, 1957 was characteristic of the calls following the Twentieth Congress to use "the best" of foreign science while remaining on guard against the importation of hostile ideology.

81. Literary Gazette, February 24, 1962; translated in CDSP, Vol. 14, No. 12, pp. 22-24.

82. In the Economics Gazette (March 26, 1962) Kapitsa argued that, had Soviet scientists listened to Soviet philosophers about cybernetics, quantum mechanics, and the theory of relativity, then there would have been no Soviet successes in space exploration

or nuclear energy. He was criticized for his pessimistic outlook about fruitful co-operation between scientists and philosophers.

83. A summary of Soviet work in medicine was in A. A. Vishnevskii's article in Pravda, January 13, 1969; translated in SC:RNI, February 1969. A good account of the problems in pedagogy was in Glushkov, et al., "Scientific Problems of Programmed Instruction and Methods of Its Implementation," in the Ukranian journal Radyanska shkola, No. 6, 1960, pp. 17-23; translated in SC:RNI, January and February 1968.

84. The problem of confusing the issue of quality-quantity with quantity-inaccuracy also was the source of many disputes in "medical cybernetics," that is, the alleged desire to replace the "intuition" of the human diagnostician with the statistical precision of computers. In the social sciences, a sophisticated article by I. V. Blaumberg and E. G. Yudin, "A Systems Approach to Social Research" (V.f., No. 9, 1967, pp. 110-117) points out that the qualitative and quantitative do not exist in an antithetical relationship.

85. D. A. Kerimov, in V. N. Kurdryavtsev, ed., Problems of Cybernetics and Law (Moscow: 1967). Translated in entirety in JPRS 43954. The author, obviously, cannot abandon the class-structure and exploitation explanation of all the negative features of U.S. society.

86. "Questions Pertaining to Investigative Reasoning in the Light of Information Theory," A. R. Ratinov in ibid.

87. A. R. Shlyakov, "The Prospects Involved in the Utilization of the Achievements of Cybernetics in the Operation of Legal Establishments," in ibid. Shlyakov replaced Kerimov as head of the Cybernetics and Law Division. The First All-Union Co-ordinating Conference was characterized by Shlyakov as only an "exchange of opinions." He also said: "Both in Moscow and Leningrad the section 'Cybernetics and Law' consisted mostly of enthusiastic cyberneticists who were carried away by ideas, and at first worked merely on the propagandization of the potentials of cybernetics in law so as to attract the attention of members of the legal profession to this problem.

88. An early reference to this system, which was openly discussed mostly in the early sixties, identified Khrushchev himself as having an interest in it. See N. Levinson, "Some Problems of Mechanizing Office Work," Izvestia, April 6, 1957, p. 2.

89. The odd coherence of the Stalinist system for science was evident in economics. Nemchinov was removed from the Timiryazev Agricultural Academy and the statistical division of the USSR Academy of Sciences for believing in the chromosome theory of heredity (V. Treml, Studies in the Soviet Union, Vol. 5, No. 2, p. 12). Kantorovich has been a very controversial figure, and may have suffered for his association with the planner N. Vosnesensky, who was executed in 1950. (G. Segal, Problems of Communism, Vol. 15, No. 2). In Stalin's 1952 work, Economic Problems of Socialism in the USSR, it was made perfectly clear that any normative contributions from economists on running the system would not be welcomed, since that was the task of the state and planning organs (see A. Becker, "Input-Output and Soviet Planning," RAND RM-3532-PR, March 1963).

90. See, for example, "Economists and Mathematicians at the Round Table," Problems of Economics, 1964, No. 9. A mild summary of this exchange appeared in the

English language magazine USSR. This summary was preceded by the comment: "The 'father' of cybernetics, Norbert Wiener, in an interview with journalist Dyson Carter, said that he envied the Soviet mathematicians who are able to apply cybernetics on a country-wide scale.

91. Later, Menchinov and others tried to legitimatize some of the work of mathematical economics and econometrics by pointing to earlier Soviet and Russian priorities in the field (for example, see A. Nove and A. Zauberman, "A Resurrected Russian Economist of 1900," Soviet Studies, Vol. 13, January 1962).

92. An excellent review of Soviet work in mathematical economics is in J. P. Hardt, ed., Mathematics and Computers in Soviet Economic Planning (New Haven: Yale University Press, 1967).

93. They may not have even understood the need for one (and only one) criterion. Gatovskii's (presumably innocent) question about the possiblity of applying several simultaneous criteria was omitted from subsequent reports of the "round table" talks (J. P. Hardt, ed., Mathematics and Computers in Soviet Economic Planning, p. 92). Glushkov, evidently in order to continue his research in the absence of a clear decision, selected minimum time (ibid., p. 91).

94. Belkin, Tretyakova, and Birman, "Cybernetics and Planning," Economics Gazette, November 16, 1960. Belkin has an interest in Soviet banking, and has argued that suboptimization should start here since the banks extend credit to enterprises and hence their maximum profit assures the best state investment (see "Reform and the Bank," Pravda, December 9, 1966, as well as "Electronics without Miracles" with Berg and Birman, Izvestia, September 9, 1966).

95. The question is whether Kantorovich's "relative scarcity" model would not bring the "supply and demand" mechanism into the Soviet economy.

96. See Berg, Birman, and Belkin, "Electronics without Miracles," and the "Round Table" discussions.

97. One planner has offered an algorithm to predict the response of administrators to his findings: if the finding increases his domain, the model is correct; if it decreases his domain, the model is incorrect.

98. I. Birman, "When Savings are Unprofitable," Izvestia, January 21, 1968.

99. See Berg and Kolman, eds., The Possible and Impossible in Cybernetics.

100. V. Glushkov, Cybernetics and Mental Labor (Knowledge Publishing House: 1965); translated in JPRS 32702.

101. "Contemporary Problems in the Organization of Management," in Applications of Electronic Computers in Production Control (Moscow: 1966); translated in SC:RNI, June 1967.

102. V. M. Glushkov, "What is Cybernetics?" Nedelia, No. 46, 1964.

103. An excellent analysis of Soviet problems with computer development and utilization is presented in Wade Holland's "Soviet Computing, 1969; a Leap into the Third Generation?," Soviet Cybernetics Review (the new name for SC:RNI), Vol. 3, No. 7, 1969, pp. 1-19.

104. We have mentioned the extent to which "economic cybernetics" and mathematical economics may have lost the stage to the "economic accountability" reforms that have followed Khrushchev's rule. Another "cybernetics controversy" in miniature may have occurred in the pages of the Literary Gazette in 1966. On May 19, N. Amosov praised cybernetics and medicine and criticized the continuing lack of effort in this direction at the Ukrainian Academy. On May 26, V. Delov charged that the careless application of cybernetics could damage the most important element of the doctor-patient relationship—its essential humanity. On June 2, A. Vishnevskii and M. Bykhovsky stated the doctor could only be helped but not replaced by computers. Finally, on June 16, N. Yudayev asserted that cybernetics had no panaceas, and the major problem was to increase the effectiveness of medical research by granting more independence and incentives to institutions and their directors.

105. Of course, such omnipotence is largely fanciful, and in the Soviet Union one can find policies that are in fact quite flexible in their implementation (for example, by simply being ignored). Ad hoc solutions to particular problems seem to occupy a prominent place in party and government decisions, where the detail of concern is so minute that individual cases simply have to be considered. The quest for "economic accountability" and economic incentives seem to have been an attempt to find a kind of controlled flexibility and limited decentralization which, of course, is heralded as "democratic centralism" after the fact. Earlier discussions about the unified state computer system, with its alleged provisions for enterprise incentive and initiative, were conducted under the same banner.

106. This tendency to counterpose individual experiences without much regard for different levels of generalization may be an echo of a populist spirit, although in the Soviet context it might be called "proletarianism." Of course, it could also be a good way to prevent the consolidation of durable "interest groups" among specialists in Soviet society.

107. The continued presence of the demand over the years probably indicates that the job of finding the "Marxist-Leninist" bases has not been done, just as the orthodox demands after the Zhdanovshchina indicated that all specialists had not fallen into line.

108. A. I. Berg, "Cybernetics in the Service of Communism," a paper in the collection by Berg and Kolman, eds., The Possible and Impossible in Cybernetics.

POSTFACE:
SCIENCE AND PUBLIC POLICY— A VIEW FROM MIT

Eugene B. Skolnikoff

The Massachusetts Institute of Technology became a major actor in national and
international affairs concerned with science and technology beginning with the early
days of World War II. This public policy involvement took on an intensity that per-
vaded many aspects of the Institute's functions and brought about substantial changes
that persisted long after the war. The pattern had been set at the Institute's founding
in 1861, when William Barton Rogers first made public service one of the major
objectives of the new institution. He could not have foreseen, however, the wartime
and postwar developments that saw MIT interacting with public affairs on such an
intensive scale.

This interaction took many different forms: creation and management of large-scale
military R&D laboratories; contract and grant relationships for basic and applied
research; conduct of major studies for the government; and not least, faculty involve-
ment in government as full-time officers while on leave or as part-time consultants.
On the national scene, the first Special Assistant to the President for Science and
Technology, Dr. James R. Killian, came from the MIT presidency. Dr. Jerome
Wiesner, MIT's current president, filled the Special Assistant's post during the
Kennedy Administration and part of the Johnson Administration.

In less formal ways, MIT's leading role as an institution devoted to teaching and
research in science and technology gave it and its faculty important functions in
society as spokesmen for the fields represented at the Institute. In one sense, trends
at MIT have become models for development and change in the entire academic community. More fundamentally, the research products of MIT's laboratories and the
graduates of its educational process, particularly in the last thirty years, have played
a substantial part in bringing about change in the society as a whole.

These developments, which have drastically altered MIT's role since 1940, have
been by no means without their costs. They imply a degree of interaction with current
affairs that inevitably complicates those aspects of a university's mission that require
calm and dissociation. They lead to a dependence on outside institutions, particularly
government, that has substantial undesirable consequences with respect to financial
stability and independence of scholarship. Further, these developments raise basic
issues with regard to responsibility for the societal effects of the application of the
knowledge generated within the Institute.

The costs of these developments, which are becoming more obvious subjects of concern today, must of course be compared to the positive side of the ledger: the steady

upgrading of the competence of research and teaching at MIT, the broadening of
fields and interests, the exciting esprit that accompanies leadership, the growth of
influence through example and innovation, and the satisfaction of performing extensive public service.

During this postwar period there were also important internal changes at the Institute, in particular in broadening the scope of its educational interests beyond the
science and engineering core. There were many changes of some importance, but
the most relevant here were the moves to create first-rate faculties in the humanities
and social sciences. The impetus dated from the early thirties under the leadership
of Karl T. Compton, but the expansion flowered under Presidents Killian and Julius
A. Stratton in the fifties and sixties. The School of Humanities and Social Sciences
received strong support within the MIT structure and was able to attract high-quality
faculty in a substantial number of fields outside the natural sciences and engineering.
The Economics and Linguistics Departments are considered at the top of their fields,
the young Political Science and Psychology Departments quickly garnered nation reputations among the best in the country. In other fields within the Humanities Department —
music, history, graphic arts — exciting faculty came to the Institute and subtly altered
the climate of what had been considered a purely "technological" institution.

All of these fields, and others outside the core of natural science and engineering,
stand on their own ground in their own disciplines. But they are also conditioned by
their setting in a predominantly technical university, so that their focus and interests
are heavily influenced, and illuminated, by the surrounding atmosphere of science and
technology. The setting and the atmosphere encouraged rigorous quantitative analysis,
or accelerated the use of scientific tools, or influenced the choice of subjects for
study, or in other ways tied the humanities and social sciences to natural science and
technology in strong and often unique ways. In addition, a host of interdepartmental
research facilities were established (for example, The Center for International Studies,
The Urban Systems Laboratory, the Center for Space Research, the Environmental
Laboratory) that provide institutional means for conducting research that crosses
natural and social science boundaries.

All of these postwar changes at MIT, in its internal development and in its ties to
the broader American community, were of course paralleled by major developments
in the United States related to science and technology. The support for science and
technology from the public purse grew to a level that projected policy for science
and technology prominently into the political arena. Over a period of a very few years,
the federal government became the main patron for basic science; a major supporter

of higher education; heavily dependent on the results of R&D; the recipient of the societal problems bequeathed by technology; and responsible for the direction, the health, and the results of the scientific and technological enterprise in the United States. The research reported in this volume attests to the new issues and relationships that were projected on the American scene since the end of World War II but of course represents only the tip of the iceberg of relevant, and now important, subjects.

It was against this general background of developments within MIT and on the national science, that the decision was taken in 1960 in the embryonic Political Science Department (then a section of the Department of Economics and Social Science) to establish a Science and Public Policy Program. By then it was clear that the growing federal responsibility for science and technology was posing increasingly important issues for public policy. And it was also clear that the pervasiveness of those subjects throughout the whole range of governmental responsibilities from foreign policy to urban policy had added a new and as yet little understood dimension to many areas of public policy. Moreover, the effects of the application of new technological developments, many of them far from the original objective of the technology, were becoming increasingly significant to society at large and usually were little understood.

MIT's central involvement and competence in science and technology, coupled with its young Social Science and Humanities Departments oriented toward the natural sciences and technology, made the decision to develop the field of science and public policy[1] an obvious one. It was not as obvious how the program should be organized or what areas should be identified as the foci for research, since the subject itself is so broad as to defy easy categorization or simple ordering of priorities.

Several choices were made in establishing the program at MIT that, for good or ill, dictated the pattern of development over the succeeding decade. Many universities now have programs of one kind or other in science and public policy, but there has proved to be little consistency among them; each was a response to the idiosyncrasies of a local situation, and to the preferences of the individuals most involved at the time. The same was true at MIT. Some of the variety will be noted later.

At MIT, the key individual was Robert C. Wood, at the time a professor of political science, who first undertook the establishment of a program, shortly after Dr. Killian's return from his White House assignment for President Eisenhower. He enlisted Dr. Killian's aid in launching one of the core courses, but it was largely Professor Wood who took the original responsibility for the overall program.

Two major choices were made. One was that the education and training of students, rather than research, was to be the primary objective, especially at the graduate level. The other was that the program would, initially at least, be a social science-based effort, using the framework of analysis of the social sciences, and especially political science, as the way into the issues. Neither of these choices could be exclusive; that is, research was obviously essential as an inseparable aspect of graduate education, and science and public policy issues require the involvement of natural scientists and engineers. Still, the education and social science emphases were central to subsequent program development.

The rationale behind these choices was important, and in part marked the difference between MIT and later programs at most other universities. The decision to emphasize teaching rather than research was partly in response to the belief that the most important product in the short run would be trained people for positions in universities and government. Moreover, it was felt that it was important to begin to expose more MIT students in all fields to a systematic analysis of the political and other societal issues surrounding science and technology. These objectives could best be achieved quickly by mounting a series of graduate and undergraduate courses, rather than first attempting to build a sizable research effort. In addition, it was recognized that much of the research that should be done would perforce be multidisciplinary in character, and could more effectively be stimulated once the social science competence had been securely established. In any case, MIT was already engaged in research that could be considered in the science and public policy field, though not within any "program" of that name or objective. It seemed more important to emphasize the coherent teaching aspects.

The decision to base the program in political science had much the same elements: a desire to build political science competence in the subject to allow more effective multidisciplinary research later and a recognition that MIT was already engaged in related research but without adequate social science inputs. Most important, there was the conviction that in fact the political implications of the new age brought about by science and technology justified a major effort by political scientists to understand those implications and to apply their skills to resolving the increasingly central political issues thus raised. In effect, part of the motivation was to influence the discipline of political science that tended to ignore these issues.

Thus, the MIT Science and Public Policy Program was established within the Political Science Department as one of the fields of emphasis of that department. A

substantial number of courses were developed over the years at both the graduate
and undergraduate levels. Originally, the effort was focused at the graduate level,
but over time both were included. Related programs were developed within the department, among them being defense studies, urban politics, and special aspects
of international relations: space, arms control, and the oceans. As expected, related activities developed at other parts of the Institute including: studies of R&D
management in the Sloan School of Management: the economics of technological innovation in the Economics Department; environmental, communications, and energy
technology issues in several departments in the School of Engineering. In some, the
ties between the Science and Public Policy Program and the research and teaching
based elsewhere at the Institute were close; for others, the projects went their own
way.

More recently, the pace of evolution for MIT as a whole has moved dramatically in
the direction of greater concern for the social and political setting in which the research at the Institute is done or in which the results of research will be applied. The
ultimate results of this new movement manifested throughout MIT are not yet clear.
So far, it has brought about several developments related in subject matter to the
Science and Public Policy Program. A spate of new courses and educational programs
has been started in a wide variety of departments or on an interdepartmental basis:
for example, courses in technology assessment, legal and social aspects of pollution,
and energy policy; and new programs such as an interdepartmental Sc.D. in Public
Systems Engineering, dealing with the regulation of technology, and an interdepartmental S.M. Program in the Social Aspects of Technology. In addition, there has
been new emphasis on developing research programs broadly concerned with science
and policy issues, and these initiatives have emerged at many diverse points in the
Institute.

The list of such developments at MIT is a long one. In fact, the trend is so striking
that in some sense it can be said that MIT is becoming an institution in which one of the
most basic concerns is with science and public policy. The existing Science and Public
Policy Program within political science is but one part of the total picture. For one
portion of MIT's science and public policy activity, the program is the central element.
For another portion, the program is but one participant in the activity. For the remainder, there may be little or no connection between the program and the activity.
The rapidity of this general development at the Institute is putting strains on the program and on the Political Science Department since the requests for meaningful participation in interdepartmental efforts are multiplying at an astonishing rate. Resource

limitations mean that only a proportion of the proposals can actually be realized as joint efforts.

One major organizational innovation is taking place at MIT in the science and public policy field. In recognition of the departmental base of the present program, and in an attempt to encourage further research in the field which must, of necessity, be multidisciplinary, a new Center for the Study of Policy Alternatives has been established. The objective of the new center is in large part to create an institutional home for policy-oriented studies in subjects related to science and technology. As noted, in addition to the Science and Public Policy Program itself there are already numerous examples of such policy research at the Institute. But the birth pangs of those that must be conducted across departmental lines are usually sufficiently traumatic as to call for an easier mechanism to bring them into being and provide continuing resources. Moreover, a certain amount of entrepreneurship is often involved at the inception stage, and to have a center with the capability to take the initiative in getting such projects started can be of great value.

Within the Science and Public Policy Program, research conducted by faculty and students in the roughly twelve years of its existence has had a small number of primary foci. The earliest focus, stemming directly from Robert Wood's interests, was on the scientific community itself, that is, the political characteristics of the community and its interaction with the body politic outside. The research encompassed in this volume stems from this research strand and is typical of other work conducted in the same area.

A second strand developed when the author of this essay joined the program in 1963, though there were earlier elements along the same lines. This second focus was on the interaction of science and technology with international affairs and with American foreign policy. It has formed one of the primary interests of the program since.

A third major strand has been the more general subject of the public management of science and technology, including within that area questions of allocation of resources, management of large-scale technological projects, and public/private sector relationships. The development of this research area was also, not surprisingly, largely the result of the arrival of an additional member of the faculty, Professor Harvey M. Sapolsky.

In addition to these primary areas there have been several individual research topics, often tied to one or another field within the department, such as defense studies, international relationships, communications, or communist studies.

Thus, in sum, the Science and Public Policy Program at MIT is a small, flexible core within the Political Science Department, a program imbedded within a department

and an institution that in many respects are themselves devoted to issues of science
and public policy broadly defined. The core program attempts to stimulate, lead,
follow, or at least be aware of the myriad of related teaching and research activities
in the rest of the university but does not attempt to provide an "umbrella" for them
all. In addition, the program takes education as its priority function, and hence is
concerned with the quality and scope of teaching in the subject, and is particularly
concerned with maintaining and developing the professional standards of this bur-
geoning field.

Roughly this same arrangement is likely to continue for some time in the future,
at least until the Center for the Study of Policy Alternatives has had a chance to
evolve. It is quite possible, however, that other institutional forms may have to be
devised if MIT as a whole continues to evolve as rapidly as it recently has toward
major demands for research and teaching in the social and political implications of
science and technology. The subject itself is so clearly of growing importance that
the institutional response may have to be more massive and more closely directed.

This pattern at MIT does not appear to be typical of what has developed in other
universities as they have moved into the area of Science and Public Policy. During
the past decade there has been growing, even explosive, realization of the significance
of the field, so that now some hundred universities in the United States alone have
identifiable courses and programs in the field. It would be of little point, even if
possible, to attempt to describe in detail the pattern of development at other universities,[2] but a few generalities can be cited.

The spectrum varies from a few courses offered in a social science department,
usually political science, to large-scale research programs essentially divorced from
an education function except through the personal involvement of faculty. Most universities seem to fall toward the middle of the spectrum, that is, with a teaching
function within a departmental structure and a separately organized research center
or program. In a few cases, the teaching function has been set up as a separate degree-
granting program outside the traditional departments, but that seems to be the exception rather than the rule. On the other hand, research centers have typically been
established on a multidisciplinary or university-wide basis.

The objectives, and the quality, of the programs vary widely, a not surprising
fact, given a new and rather inchoate field that relates to so many other subjects at
once. In fact, there is a continuing discussion as to what the boundaries of the subject are and whether it should be considered a separate "field" or "discipline" at all.

In 1968, in recognition of the growing number of university activities in the area,
a seminar was organized at the American Association for the Advancement of Science

(AAAS) annual meeting to explore the state of the field and to determine whether any concrete organizational steps should be taken on a multiuniversity basis. The result of that meeting was the creation of the Science and Public Policy Studies Group, a loose affiliation of universities and individuals interested in developing the field. The objectives of the Studies Group were modest: essentially to provide a center of communication and information to serve affiliates, to arrange seminars, and to provide an information clearinghouse. The Studies Group was established at MIT, with the author as chairman and a small headquarters facility.

After four years of existence, the Studies Group had grown to the point where over 500 individuals had asked to join and some 100 universities in the United States were affiliated with it. There were also a substantial number of foreign correspondents. The Newsletter proved to be an important source of information and communication, and other activities have been well attended and received.

Increasingly, however, the leadership of the Studies Group became concerned with the question of its future structure and role. The decline in government support for science and technology, the public questioning of the ultimate value of technology, and the growing crisis of values within science and engineering themselves, all emphasize the importance of research and teaching in the field. But should science and public policy be considered a discipline, as some argue, and should the Studies Group be the nucleus of the professional association of that discipline: or should the Studies Group consider itself as simply one means of stimulating activities in the field without trying to circumscribe the subject with boundaries that may result in creating barriers rather than communication?

A different level of choice is also pertinent. The objectives of the Studies Group were essentially passive in the sense of being limited to a service function for its affiliates. Is this enough, given the size of the network that has been established, and given the needs of society and of government to understand the problems they must face with relation to science and technology? Should, for example, the Studies Group network be used as a means of organizing studies on a national basis, of mobilizing groups to serve government, of attempting to set research priorities in the field, or even of developing an independent critic of government policy? And what of the possibility of cross-national action?

A step toward the resolution of these issues was taken in December 1972, when the Studies Group formally metamorphosed from an independent, incorporated body into a new Committee on Science and Public Policy of the American Association for the Advancement of Science (AAAS). This committee, whose initial membership is drawn largely from the steering committee of the former Studies Group, intends to

continue some of the "trade association" activities of the Studies Group. By virtue
of the much higher level of financial and staff support available through AAAS, however, a much broader range of potential functions is opened up. The Newsletter,
which has proved its value as an independent publication, continues at MIT under the
title Public Science.

Whatever route the new committee chooses to take, it is clear that the field is
becoming an important focus of research and training in American universities. The
MIT program was one of the earliest formal programs, but now a substantial number
of others have emerged and a healthy body of research is being developed. Though
the field will undoubtedly always be somewhat amorphous—somewhat akin to area
studies rather than a separate discipline—it is bound to grow in significance in response to the growing importance of the subject matter to society. It is essential that
the research and the training reflect that reality in quality and scope.

NOTES

1. The choice of name for the field may have been unfortunate—it is meant to include
both science and technology and their interaction with public policy and public affairs.

2. Descriptions of some programs are given in Lynton K. Caldwell, ed., Science,
Technology, and Public Policy: A Selected and Annotated Bibliography, Vol. 1 (Bloom-
ington, Indiana: Department of Government, Indiana University, 1968); in various
issues of Public Science (formerly the sppsg Newsletter), published by the MIT Press;
and in Teaching and Research in the Field of Science Policy—A Survey, a staff study
for the Subcommittee on Science, Research, and Development of the Committee on
Science and Astronautics, U.S. House of Representatives, 92nd Congress, 2nd Session
(Washington: U.S. Government Printing Office, 1973).